Math Kangaroo USA
Problems and Solutions

Grades 7 & 8

**Odd Years
2007–2025**

Editor in Chief

Agata Gazal
Chief Editorial Officer for Math Kangaroo USA
Billings, MT

Reviewers and Contributors

Joanna Matthiesen
Chief Executive Officer for Math Kangaroo USA
Granger, IN

Izabela Szpiech
Chief Financial Officer for Math Kangaroo USA
Chicago, IL

Kasia Nalaskowska
Chief Information Officer for Math Kangaroo USA
Aurora, IL

Magdalena Teodorowicz
Chief Design Officer for Math Kangaroo USA
Cordova, TN

Professor Andrzej Zarach, Ph.D.
Math Content Reviewer
East Stroudsburg University, East Stroudsburg, PA

Book Design

Jossea K. Rilea
Designer at LX Design Lab
Saratoga Springs, NY

We would like to give special thanks to countless other people who contributed to the problems and solutions for this book since 1998. Primarily, a big thank you to the Math Kangaroo question writers from all over the world that are part of the AKSF organization (www.aksf.org). Math Kangaroo solution writers also include Math Kangaroo USA competition organizers and Math Kangaroo alumni. We would also like to thank the hundreds of educators who gave us feedback on the questions and solutions and finally the hundreds of thousands of students who took the Kangaroo challenge over the last two decades. Thank you all for your help in developing this book.

Copyright © 2025 by Math Kangaroo USA, NFP, Inc.
www.mathkangaroo.org

For additional copies of this book, please contact the publisher:
Math Kangaroo USA
info@mathkangaroo.org

ISBN: 9798989988372

Preface

Welcome to a world of math challenges and exciting problem-solving! Whether you're a student looking to sharpen your skills or a teacher eager to inspire young minds, this book is a treasure trove of stimulating questions and solutions.

Inside, you'll find 300 captivating problems, drawn from 10 years of the Math Kangaroo Competition (2007-2025 odd years), designed for 7th- and 8th-grade students. These questions are carefully selected at the annual Kangourou sans Frontières meeting, where mathematicians from over 100 countries gather to ensure each problem is engaging and age-appropriate. Each test consists of 30 questions, categorized by difficulty—easy, medium, and difficult—so that all students can find their challenge level.

This easy-to-use resource book is more than just a collection of problems. It's a journey into the world of math and logic, with visually appealing questions and insightful solutions that encourage children to think critically about the world around them. Problem-solving is a skill students use daily, often without realizing it. This book is designed to help them practice logical reasoning, enhance their math skills, and reflect on their problem-solving process.

We hope this book will inspire not only students who have a passion for math, but also educators who love to bring unconventional, thought-provoking challenges into the classroom. Whether you're seeking to improve mathematical thinking or simply enjoy the thrill of solving puzzles, we believe this book will provide both fun and valuable learning experiences.

Enjoy the journey!

Joanna Matthiesen, CEO

Color Key

Each test has 30 questions with 3 levels of difficulty

GREEN	YELLOW	RED
Easy	Medium	Difficult
Q 1-10	Q 11-20	Q 21-30
3 Points	4 Points	5 Points

TABLE OF CONTENTS

Part I PROBLEMS 7
2007 9
2009 15
2011 21
2013 27
2015 33
2017 39
2019 45
2021 51
2023 57
2025 63

Part II SOLUTIONS 71
2007 73
2009 79
2011 85
2013 91
2015 97
2017 103
2019 109
2021 115
2023 123
2025 131

Part III ANSWER KEY 139

Part I
Problems

2007

2007

3 Points Each

1 $\dfrac{2007}{2+0+0+7}$

(A) 1003
(B) 75
(C) 223
(D) 213
(E) 123

2 Rose bushes are planted in a line on both sides of a path. The distance between the bushes is 2 m. What is the largest number of bushes that can be planted if the path is 20 m long?

(A) 22
(B) 20
(C) 12
(D) 11
(E) 10

3 The distance ran in a marathon is 26.2 miles. Jerry started the marathon at 1:37 p.m., and he reached the finish line at 4:18 p.m. How many minutes did it take him to finish the marathon?

(A) 131
(B) 91
(C) 151
(D) 185
(E) 161

4 What is the sum of the number of dots on the faces we cannot see of the dice shown in the illustration?

(A) 15
(B) 12
(C) 7
(D) 27
(E) another answer

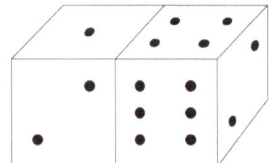

5 In the illustration, there are six identical circles inside a rectangle. The circles touch the sides of a large rectangle and each other as well. The vertices of the small rectangle lie at the centers of four of the circles. The perimeter of the small rectangle is 60 cm. What is the perimeter of the large rectangle?

(A) 90 cm
(B) 140 cm
(C) 120 cm
(D) 100 cm
(E) 80 cm

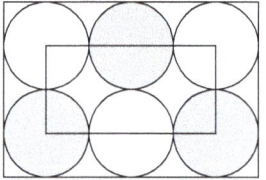

6 The points $A = (6, 7)$, $B = (7, 6)$, $C = (-6, -7)$, $D = (7, -7)$ and $E = (7, -6)$ are marked on a coordinate grid. Which line segment is parallel to the x-axis?

(A) $\overline{AD}$
(B) $\overline{BE}$
(C) $\overline{BC}$
(D) $\overline{CD}$
(E) $\overline{AB}$

7 A small square is inscribed in a big square as shown in the diagram. What is the area of the small square?

(A) 16
(B) 28
(C) 34
(D) 36
(E) other

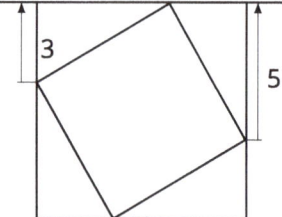

8 What is the smallest prime number that divides the sum $3^{11} + 5^{13}$?

(A) 2
(B) 3
(C) 5
(D) $3^{11} + 5^{13}$
(E) 1

9 A palindromic number is one that reads the same backwards as forwards; for example, 13931 is a palindromic number. What is the difference between the largest 6-digit palindromic number and the smallest 5-digit palindromic number?

(A) 989989
(B) 989998
(C) 998998
(D) 999898
(E) 899998

10 What is the solution of this equation: $2^{2007} \cdot x = 2^{2006}$?

(A) 1
(B) 2
(C) $\frac{1}{2}$
(D) 2^2
(E) 2^{2008}

4 Points Each

11 x is a strictly negative integer. Which of the expressions is the greatest?

(A) $x + 1$
(B) $2x$
(C) $-2x$
(D) $6x + 2$
(E) $x - 2$

12 The squares shown in the diagram are formed by sections of segment AB (AB = 24 cm) and by the line segments $AA_1A_2A_3 \ldots A_{11}A_{12}B$. Find the length of $AA_1A_2A_3 \ldots A_{11}A_{12}B$.

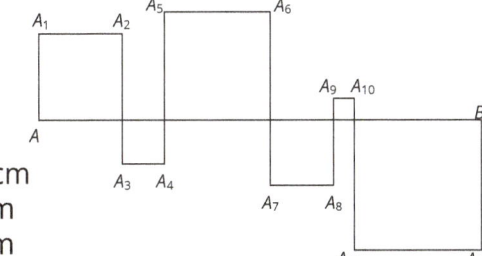

(A) 106 cm
(B) 96 cm
(C) 72 cm
(D) 56 cm
(E) 48 cm

13 Six points were selected on the parallel lines a and b: 4 on line a and 2 on line b. What is the total number of triangles with vertices at the given points?

(A) 6
(B) 8
(C) 12
(D) 16
(E) 18

14 A mechanical kangaroo starts walking in the grid starting at square A2 in the direction shown by the arrow, as shown in the diagram. It can only go forward, jumping in a single jump from the center of a square to the center of an adjacent square (adjacent squares have a common side). The kangaroo cannot jump out off of the grid, and it cannot go into a shaded square. If it cannot jump forward, it turns 90° right and then jumps. The kangaroo will stop when it cannot go forward after turning right. On which square will it stop?

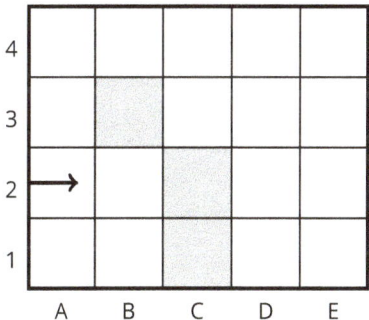

(A) B2
(B) A1
(C) E1
(D) D1
(E) It will never stop.

15 To obtain 9^9, to which power does 3^3 need to be raised?

(A) 2
(B) 3
(C) 6
(D) 19
(E) 18

16 Triangles *ABC* and *CDE* are equilateral and congruent. If the measure of angle ∠*ACD* = 80°, what is the measure of angle ∠*ABD*?

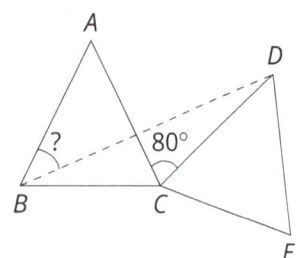

(A) 25°
(B) 30°
(C) 35°
(D) 40°
(E) 45°

---

17 What percent of the elements of the set of natural numbers {1, 2, 3, 4, . . . , 10000} are squares of natural numbers?

(A) 1%
(B) 5%
(C) 10%
(D) 50%
(E) 0.1%

---

18 Segments *OA, OB, OC,* and *OD* are drawn from the center *O* of square *KLMN* to its sides so that *OA* ⊥ *OB* and *OC* ⊥ *OD* (as shown in the figure). If the length of the side of the square equals 2, the area of the shaded part equals:

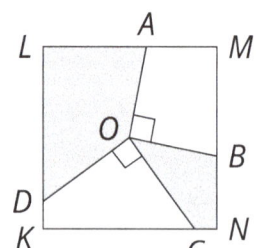

(A) 1
(B) 2
(C) 2.5
(D) 2.25
(E) It depends on the choice of points *B* and *C*.

---

19 Select three numbers from the grid shown in such a way that no two of the numbers are in the same row or in the same column. What is the largest sum of the numbers chosen in this way?

(A) 12
(B) 15
(C) 18
(D) 21
(E) another value

1	2	3
4	5	6
7	8	9

---

20 What is the smallest number of little squares that we need to shade in the illustration below so that the figure has an axis of symmetry?

(A) 4
(B) 6
(C) 5
(D) 2
(E) 3

5 Points Each

21 A certain broken calculator does not display the digit 1. For example, if we type in the number 3131, only the number 33 is displayed, with no spaces. Mike typed a 6-digit number into that calculator, but only 2007 appeared on the display. How many different numbers could Mike have typed?

(A) 12
(B) 13
(C) 14
(D) 15
(E) 16

---

22 A man took a 2-hour walking tour. At first, the road was level, and then he had to climb uphill. Going back he went downhill at first, and then he was on level ground again. The whole trip took him 2 hours. How many kilometers did he travel if his speed was 4 km/hr on the level part, 3 km/hr when he was going uphill, and 6 km/hr when he was going downhill?

(A) Not enough information is given.
(B) 6 km
(C) 7.5 km
(D) 8 km
(E) 10 km

23 To draw a table with 12 cells, 5 horizontal and 4 vertical lines were used (shown in the figure). If we use 6 horizontal lines and 3 vertical lines we will get a table with 10 cells. What is the greatest number of cells that can be obtained using 15 lines to draw the table?

(A) 56
(B) 27
(C) 32
(D) 40
(E) 42

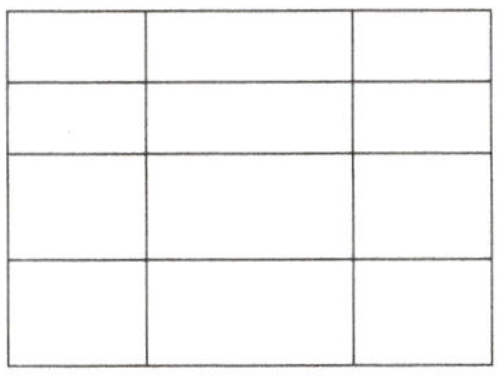

24 Segment AK is the bisector of angle A in triangle ABC. AK divides triangle ABC into two triangles with the same area. Triangle ABC is definitely:

(A) equilateral
(B) isosceles
(C) right
(D) acute
(E) obtuse

25 A positive integer n has 2 natural divisors, while $n + 1$ has 3 natural divisors. How many natural divisors does $n + 2$ have?

(A) 2
(B) 3
(C) 4
(D) 5
(E) It depends on the value of n.

26 A 3 × 3 table contains natural numbers (see the illustration). Nick and Pete crossed out four numbers each in such a way that the sum of the numbers crossed out by Nick is three times as great as the sum of the numbers crossed out by Pete. The number which remains in the table is:

(A) 4
(B) 7
(C) 14
(D) 23
(E) 24

4	12	8
13	24	14
7	5	23

27 Five integers are written around a circle in such a way that no two numbers next to each other have a sum divisible by 3 and the sum of the three other numbers is not divisible by 3. Among those 5 numbers, how many are divisible by 3?

(A) 0
(B) 1
(C) 2
(D) 3
(E) It is impossible to determine.

28 The illustration shows a square tile. The curved lines are quarters of a circle with a radius equal to half of the side of the tile. The length of the curved lines on one tile is 5 dm. We build a square using 16 tiles. What is the maximum length of a continuous line made by the quarters of a circle (the curves)?

(A) 75 dm
(B) 100 dm
(C) 105 dm
(D) 110 dm
(E) 80 dm

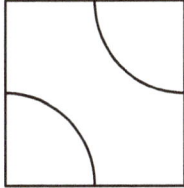

29 How many three-digit numbers divisible by 9 have the following property: the sum of the digits of the quotient of the number divided by 9 is 9 less than the sum of the digits of the original three-digit number?

(A) 1
(B) 2
(C) 4
(D) 5
(E) 11

---

30 A Kangaroo calculator can only do the following operations: multiply a number by 2 or by 3, or raise it to the 2nd or the 3rd power. If we start with the number 15, which of the following values can be obtained by using this calculator 5 times consecutively?

(A) $2^8 \cdot 3^5 \cdot 5^6$
(B) $2^8 \cdot 3^4 \cdot 5^2$
(C) $2^3 \cdot 3^3 \cdot 5^3$
(D) $2^6 \cdot 3^6 \cdot 5^4$
(E) $2 \cdot 3^2 \cdot 5^6$

2009

2009

3 Points Each

1 Which of these numbers is the greatest?

(A) 2009
(B) 2 + 0 + 0 + 9
(C) 200 − 9
(D) 200 · 9
(E) 200 + 9

2 Four boys and four girls attended Adam's party. The boys danced only with the girls and the girls danced only with the boys. Afterward, when asked the question, "With how many different people did you dance?" the four boys answered: 3, 1, 2, 2, while three of the girls answered: 2, 2, 2. With how many boys did the fourth girl dance?

(A) 0
(B) 1
(C) 2
(D) 3
(E) 4

3 The star in the illustration is constructed from 12 identical equilateral triangles and has a perimeter of 36 cm. What is the perimeter of the shaded hexagon?

(A) 6 cm
(B) 12 cm
(C) 18 cm
(D) 24 cm
(E) 30 cm

4 While preparing for the Math Kangaroo contest, Jack decided to solve one question from each odd numbered page in his collection of problems. He began on page 15 and finished on page 53. How many problems did Jack solve?

(A) 19
(B) 20
(C) 27
(D) 38
(E) 53

5 The large square has an area of 1 and has been divided into smaller squares, as shown. What is the area of the small black square?

(A) $\frac{1}{18}$
(B) $\frac{1}{108}$
(C) $\frac{1}{162}$
(D) $\frac{1}{324}$
(E) $\frac{1}{1000}$

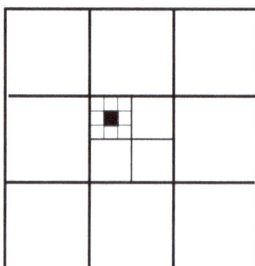

6 The product of four different positive integers equals 100. Their sum is equal to:

(A) 10
(B) 12
(C) 15
(D) 18
(E) 20

7 Some dogs and cats are playing together. The number of cat paws is twice the number of dog noses. The number of cats must be:

(A) twice the number of dogs.
(B) equal to the number of dogs.
(C) equal to half the number of dogs.
(D) $\frac{1}{4}$ the number of dogs.
(E) $\frac{1}{6}$ the number of dogs.

---

8 In the diagram the points Q, S, R are collinear, $\angle QPS = 12°$ and $|PQ| = |PS| = |RS|$. The measure of $\angle QPR$ is equal to:

(A) 36°
(B) 42°
(C) 54°
(D) 60°
(E) 84°

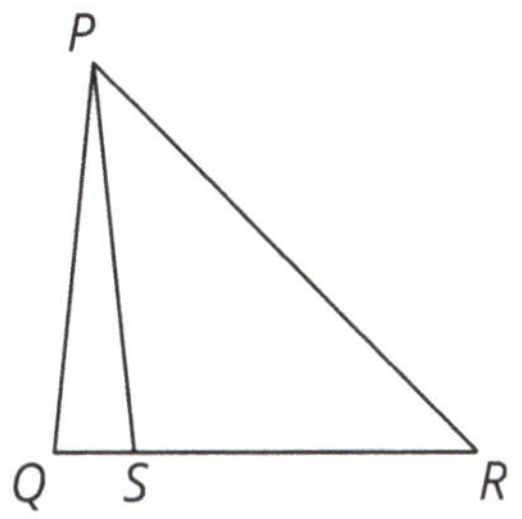

---

9 Mother made enough juice to entirely fill either 12 large jars or 20 smaller jars. She already filled 9 of the larger jars when she decided to pour the remaining juice into the smaller jars. How many of the smaller jars will Mother need?

(A) 3
(B) 4
(C) 5
(D) 6
(E) 8

---

10 A pair of integers is called "good" if their sum is equal to their product. How many "good" pairs of integers are there?

(A) 1
(B) 2
(C) 3
(D) 5
(E) Infinitely many.

---

4 Points Each

11 For how many positive integers is the number of digits in the base ten representation of their square the same as that of their cube?

(A) 0
(B) 3
(C) 4
(D) 9
(E) Infinitely many.

---

12 What is the minimum number of dots that need to be removed from the given figure so that no three of the remaining dots are collinear?

(A) 3
(B) 4
(C) 2
(D) 7
(E) 1

---

13 Mark measured every angle in two triangles — one acute and one obtuse. He remembered the angle measures of four of the angles: 120°, 80°, 55°, 10°. What was the measure of the smallest angle in the acute triangle?

(A) 5°
(B) 10°
(C) 45°
(D) 55°
(E) It is impossible to determine.

---

14 What fraction of the largest square is the shaded region?

(A) $\frac{1}{4}$
(B) $\frac{\pi}{12}$
(C) $\frac{\pi+2}{16}$
(D) $\frac{\pi}{4}$
(E) $\frac{1}{3}$

15 In each cell of a 10 × 19 table we write either 0 or 1. Each row and each column is labeled with the sum of its elements. The largest number of different labels that may be obtained in this way is:

(A) 9
(B) 10
(C) 15
(D) 19
(E) 29

16 The surface of the solid in the diagram consists of six triangular faces. Each vertex has been assigned a number in such a way that the sum of the numbers on the vertices of a face is the same for every face. The numbers 3 and 6 are assigned as shown. What is the sum of all the numbers assigned to the vertices?

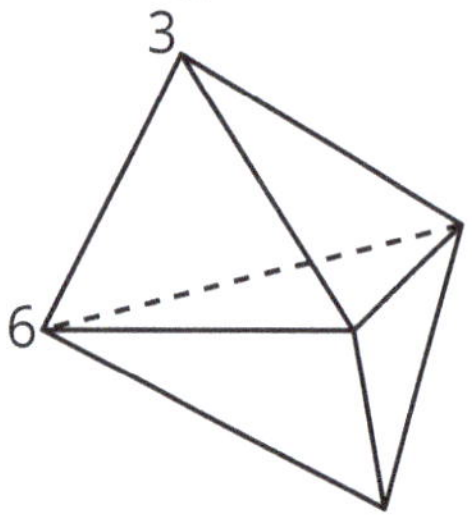

(A) 9
(B) 12
(C) 17
(D) 18
(E) 24

17 In the equation $\frac{E \cdot I \cdot G \cdot H \cdot T}{F \cdot O \cdot U \cdot R} = T \cdot W \cdot O$
the dot represents multiplication and each letter corresponds to a different digit with the same letters having the same value. How many different values can product $T \cdot H \cdot R \cdot E \cdot E$ have?

(A) 1
(B) 2
(C) 3
(D) 4
(E) 5

18 The fractions 1/3 and 1/5 are marked on the number line. Which letter corresponds to the fraction 1/4?

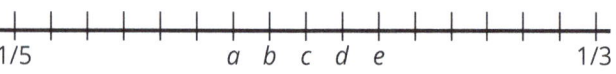

(A) a
(B) b
(C) c
(D) d
(E) e

19 The figure in the diagram is a regular nonagon (a regular 9-sided polygon). What is the measure of angle a?

(A) 40°
(B) 45°
(C) 50°
(D) 55°
(E) 60°

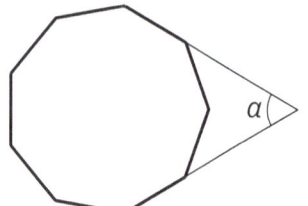

20 We form a pattern as shown in the illustration. How many of the small white tiles are necessary to build the tenth element in this pattern?

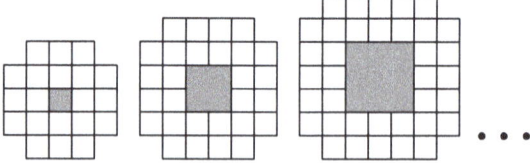

(A) 76
(B) 80
(C) 84
(D) 92
(E) 100

5 Points Each

21 An island is inhabited by two types of people: truth-tellers and liars. The truth-tellers always speak the truth and the liars always lie. 25 of the island's inhabitants stood in a line. Each of them with the exception of the first person said: "The person directly in front of me is a liar," while the person standing first in the line said: "Everyone standing behind me is a liar." How many liars stood in the line?

(A) 24
(B) 13
(C) 12
(D) 0
(E) It is impossible to determine.

22 How many ten-digit numbers can be constructed using the digits 1, 2, and 3 so that any two consecutive digits differ by exactly 1?

(A) 16
(B) 32
(C) 64
(D) 80
(E) 100

23 Consider pairs of positive integers with sum no larger than 103 and quotient smaller than $\frac{1}{3}$. The largest possible quotient of any such pair is:

(A) $\frac{27}{77}$
(B) $\frac{26}{77}$
(C) $\frac{25}{76}$
(D) $\frac{25}{77}$
(E) $\frac{26}{75}$

24 A large cube has been cut into eight rectangular prisms as shown in the diagram. What is the ratio of the total surface area of the eight pieces to the surface area of the cube?

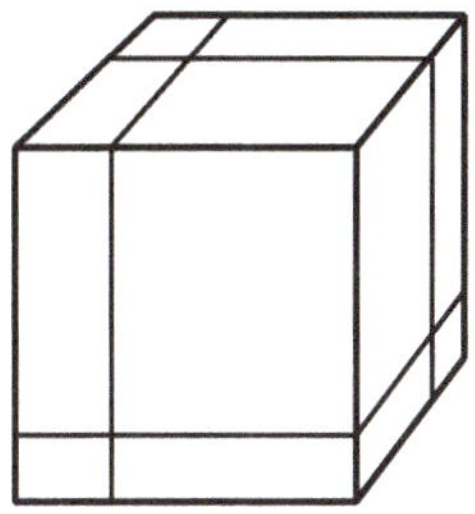

(A) 1:1
(B) 4:3
(C) 3:2
(D) 2:1
(E) 4:1

25 A divisor of a natural number is called "proper" if it is greater than 1 and smaller than the given number. For how many natural numbers is the largest "proper" divisor 45 times greater than the smallest "proper" divisor?

(A) 0
(B) 1
(C) 2
(D) 3
(E) More than 3.

26 A large square has been divided into 2009 smaller squares whose side lengths are integers. What is the smallest possible side length of the large square?

(A) 44
(B) 45
(C) 46
(D) 503
(E) It is impossible to divide a square into 2009 such squares.

27 In quadrilateral PQRS, |PQ| = 2006, |QR| = 2008, |RS| = 2007, and |SP| = 2009. Which vertices correspond to interior angles whose measure must be less than 180°?

(A) P, Q, R
(B) Q, R, S
(C) P, Q, S
(D) P, R, S
(E) P, Q, R, S

28 Consider a square measuring 6 cm × 6 cm and a triangle. If the square is placed on top of the triangle then it covers 60% of the triangle's surface area. If the triangle is placed on top of the square, then it covers $\frac{2}{3}$ of the surface of the square. The area of the triangle measures:

(A) $22\frac{4}{5}$ cm²
(B) 24 cm²
(C) 36 cm²
(D) 40 cm²
(E) 60 cm²

29 We wish to color the square grid in the diagram with colors A, B, C, and D in such a manner that no two squares with a common vertex or edge are of the same color. Some squares are already colored. What are the valid colorings of the shaded square?

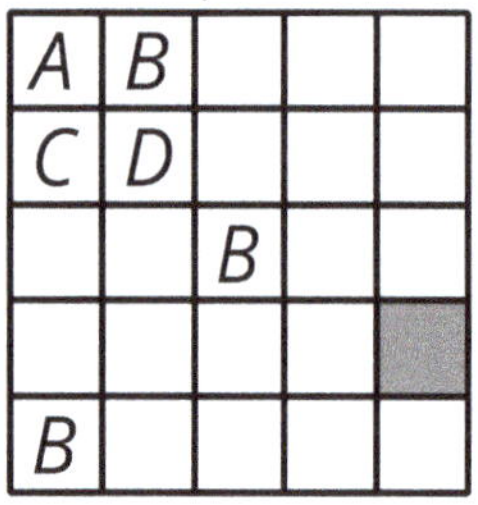

(A) Only B
(B) Only C
(C) Only D
(D) Either C or D
(E) A valid coloring does not exist.

30 In △ABC the measure of angle B is 20° and the measure of angle C is 40°. Moreover, the length of the bisector of angle A is equal to 2. What is the value of |BC| − |AB|?

(A) 1
(B) 1.5
(C) 2
(D) 4
(E) It is impossible to determine.

2011

2011

3 Points Each

1 The largest of the following numbers is

(A) 2011^1
(B) 1^{2011}
(C) 1×2011
(D) $1 + 2011$
(E) $1 \div 2011$

2 Ellie has 5 cubes and 3 tetrahedrons. How many sides do these solids have all together?

(A) 42
(B) 48
(C) 50
(D) 52
(E) 56

3 A pedestrian crosswalk consists of alternating white and black stripes. The crosswalk begins and ends with a white stripe. How wide is the crosswalk, if every stripe is 50 cm wide and there are 8 white stripes in all?

(A) 7 m
(B) 7.5 m
(C) 8 m
(D) 8.5 m
(E) 9 m

4 A certain broken calculator divides instead of multiplying and subtracts instead of adding. I entered the expression (12 × 3) + (4 × 2). What answer will the broken calculator display?

(A) 2
(B) 6
(C) 12
(D) 28
(E) 38

5 At exactly 20:11, Peter looked at his digital watch, which was set to display time in 24-hour mode. (This is also knows as military time. Note that this is 8:11 p.m. in 12-hour mode). What is the shortest amount of time, in minutes, after which the displayed time will again contain the digits 0, 1, 1, and 2?

(A) 40
(B) 45
(C) 50
(D) 55
(E) 60

6 The diagram below shows three squares: large, medium, and small. The vertices of the medium square lie on the midpoints of the sides of the large square, and the vertices of the small square lie on the midpoints of the sides of the medium square. The area of the small square is 6 cm². What is the difference between the area of the large square and the area of the medium square?

(A) 6 cm²
(B) 9 cm²
(C) 12 cm²
(D) 15 cm²
(E) 18 cm²

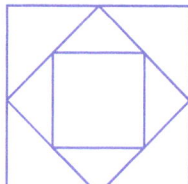

7 There are 17 houses on the street on which I live. The houses on the right side of the street are numbered with consecutive even numbers starting with 2, and the houses on the left side of the street are numbered with consecutive odd numbers starting with 1. I live in the last house on the right side of the street and my house number is 12. What is the number of the last house on the left side of my street?

(A) 5
(B) 11
(C) 13
(D) 17
(E) 21

8 All four-digit numbers the sum of whose digits is 4 were arranged in order from least to greatest. In which position is the number 2011? (Note: A four-digit number cannot begin with a 0.)

(A) 6
(B) 9
(C) 12
(D) 15
(E) 18

9 Numbers a, b, and c have the following properties: the arithmetic mean of a and b is equal to 17, and the arithmetic mean of a, b, and c is equal to 15. The number c is equal to

(A) 15
(B) 12
(C) 11
(D) 10
(E) 9

10 The sum of the smallest three-digit numbers whose digits add up to 8 and the largest three-digit numbers whose digits add up to 8 is equal to

(A) 707
(B) 907
(C) 916
(D) 1000
(E) 1001

4 Points Each

11 $\dfrac{2011 \times 2.011}{201.1 \times 20.11} =$

(A) 0.01
(B) 0.1
(C) 1
(D) 10
(E) 100

12 During three consecutive games the soccer team FC Kangaroo made a total of three goals and had one goal scored against them. This soccer team won one of these three games, lost one game, and tied one game. What was the final score of the game that FC Kangaroo won?

(A) 2–0
(B) 0–1
(C) 1–0
(D) 2–1
(E) 3–0

13 A jeweler had 9 pearls which weighed 1 g, 2 g, 3 g, 4 g, 5 g, 6 g, 7 g, 8 g, and 9 g, respectively. He made 4 necklaces and in each necklace set 2 of these pearls. The pearls in each necklace weighed 17 g, 13 g, 7 g, and 5 g, respectively. What is the weight of the pearl that was not set in any of the necklaces?

(A) 1 g
(B) 2 g
(C) 3 g
(D) 4 g
(E) 5 g

14 The circles in the map of the labyrinth below represent gold coins. An adventurer is going through a labyrinth in search of gold coins, but he may not visit any portion of the labyrinth more than once. What is the largest possible number of gold coins the adventurer can collect?

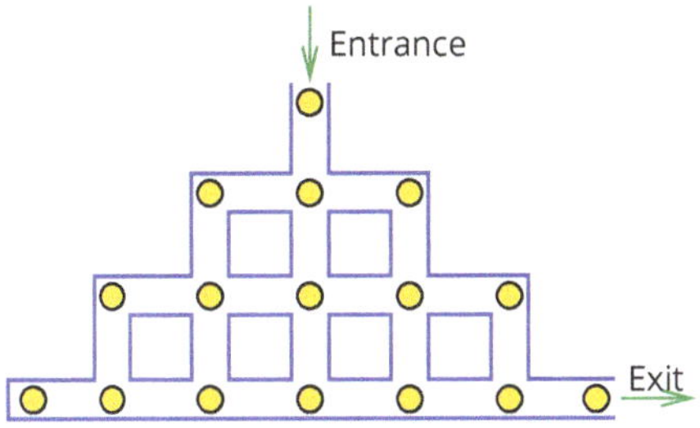

(A) 12
(B) 13
(C) 14
(D) 15
(E) 16

15 Each region in the illustration must be colored in such a way that neighboring regions are of different colors. The available colors are red (R), green (G), blue (B), and orange (O). Three regions have already been colored, as indicated by the letters in the illustration. What will be the color of region X?

(A) red
(B) blue
(C) green
(D) orange
(E) It cannot be determined.

16 Peter, who is an avid fisherman, caught 12 fish over three consecutive days. On each day, except the first day, he caught more fish than on the previous day. On the third day he caught fewer fish than the total number from the previous two days. How many fish did Peter catch on the third day?

(A) 5
(B) 6
(C) 7
(D) 8
(E) 9

17 Which two numbers can be removed from the numbers 17, 13, 5, 10, 14, 9, 12, and 16, so that the arithmetic mean remains unchanged?

(A) 12 and 17
(B) 5 and 17
(C) 9 and 16
(D) 10 and 12
(E) 14 and 10

18 Consider a plane containing the line segment DE of length 2. How many points F in this plane have the property that the triangle DEF is a right triangle with an area equal to 1?

(A) 2
(B) 4
(C) 6
(D) 8
(E) 10

19 The number a is positive and less than 1, and the number b is greater than 1. Which of the following numbers is the greatest?

(A) $a \times b$
(B) $a + b$
(C) $a \div b$
(D) b
(E) $b - a$

20 Each of the four squares in the illustration is identical. We draw one more such square in such a way that the resulting figure has an axis of symmetry. In how many ways can this be done?

(A) 1
(B) 2
(C) 3
(D) 5
(E) 6

5 Points Each

21 A square piece of paper was cut into six rectangles as shown. The sum of the perimeters of these six rectangles is equal to 120 cm. What is the area of the sheet of paper?

(A) 48 cm²
(B) 64 cm²
(C) 110.25 cm²
(D) 144 cm²
(E) 256 cm²

22 The five-digit natural number 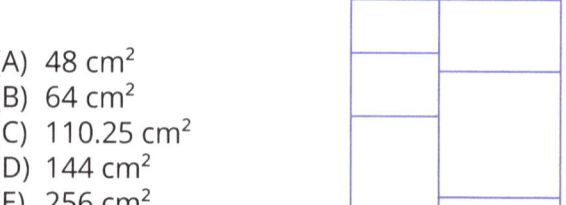 is divisible by 4, 5, and 9. The sum of the digits x and y is equal to

(A) 13
(B) 10
(C) 9
(D) 5
(E) 4

23 Mary placed two dark shapes, each made of five identical squares, on a square grid. Which of the following shapes, each also made of five squares, can be placed in the unshaded region of the grid in such a way that there will be no room left to place any of the remaining four shapes? Any shape may be rotated or turned over before it is placed in the grid.

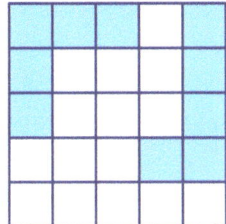

(A)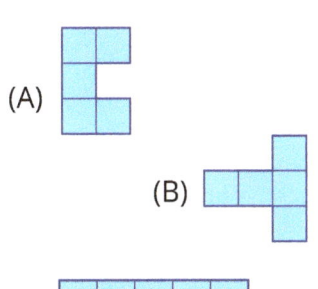

(B)

(C)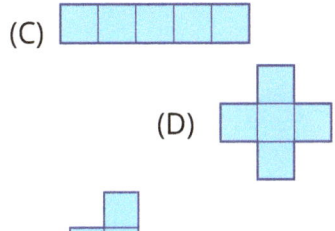

(D)

(E)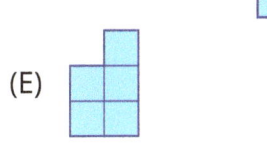

---

24 Inside a square with a side length equal to 7 cm there is a square with a side length equal to 3 cm. A third square with a side length equal to 5 cm intersects each of the other two squares (see illustration). The difference between the area of the black region and the sum of the areas of the green region is

(A) 0 cm²
(B) 10 cm²
(C) 11 cm²
(D) 15 cm²
(E) It is impossible to determine.

---

25 During target practice Bob can earn 5, 8, or 10 points for hitting the target. He hit the 10 point mark as many times as he hit the 8 point mark. Altogether, Bob managed to earn 99 points while missing the target 25% of the time. How many times did Bob fire at the target?

(A) 10
(B) 12
(C) 16
(D) 20
(E) 24

---

26 In the convex quadrilateral $ABCD$, $|AB| = |AC|$. In addition, $|\angle BAD| = 80°$, $|\angle ABC| = 75°$, and $|\angle ADC| = 65°$ (see illustration). What is the measure of $\angle BDC$?

(A) 10°
(B) 15°
(C) 20°
(D) 30°
(E) 45°

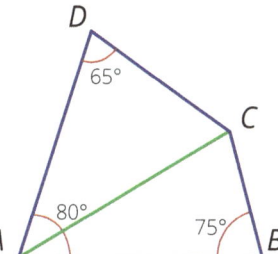

---

27 The illustration shows a cube and its layout with face $ABCD$ labeled. The cube is cut along the green line into two identical solids. Which of the following diagrams represents the layout of the cube with the bold line properly marked?

(A)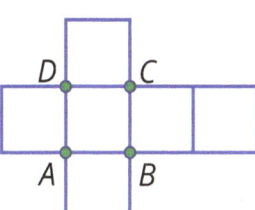

(B)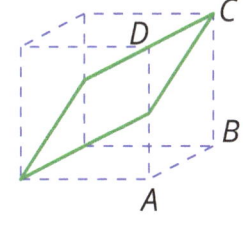

(C)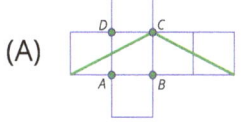

(D)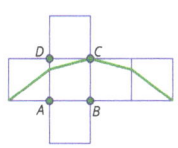

(E)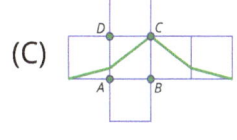

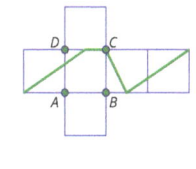

28 Each letter in the expression shown below represents a digit other than zero. Different letters represent different digits and the same letter always represents the same digit. What is the smallest whole number that can be the value of this expression?

$$\frac{K \times A \times N \times G \times A \times R \times O \times O}{G \times A \times M \times E}$$

(A) 1
(B) 2
(C) 3
(D) 5
(E) 7

29 The shape in figure 1 is made of two rectangles. Two of the side lengths are given. The shape was cut along the dotted lines and rearranged into the triangle in figure 2. The length of side *x* is equal to

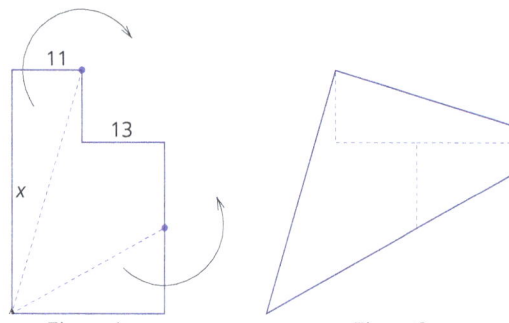

Figure 1   Figure 2

(A) 36
(B) 37
(C) 38
(D) 39
(E) 40

30 Each of three boys, Adam, John, and Karl, made a statement, as follows:

Adam: "The distance between John and me is greater than twice the distance between Karl and me."

John: "The distance between Karl and me is greater than twice the distance between Adam and me."

Karl: "The distance between John and me is greater than twice the distance between Adam and me."

At least two of these statements are true. Which of the boys is telling a lie?

(A) Adam
(B) John
(C) Karl
(D) None of them.
(E) It is impossible to determine.

2013

2013

3 Points Each

1 In the illustration, the big triangle is equilateral and has an area of 9. The lines are parallel to the sides and divide the sides into three equal parts. What is the area of the shaded part?

(A) 1
(B) 4
(C) 5
(D) 6
(E) 7

2 It is true that $\frac{1111}{101} = 11$. What is the value of $\frac{3333}{101} + \frac{6666}{303}$?

(A) 5
(B) 9
(C) 11
(D) 55
(E) 99

3 The masses of salt and fresh water in sea water in Protaras are in the ratio 7 : 193. How many kilograms of salt are there in 1000 kg of sea water?

(A) 35
(B) 186
(C) 193
(D) 200
(E) 350

4 Ann has the square sheet of paper shown below. By cutting along the lines of the square, she cuts out copies of the shape shown further to the right. What is the smallest possible number of cells remaining?

(A) 0
(B) 2
(C) 4
(D) 6
(E) 8

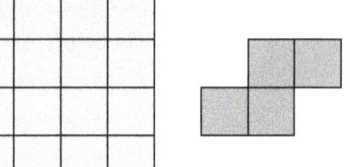

5 Roo wants to tell Kanga a number with the product of its digits equal to 24. What is the sum of the digits of the smallest number that Roo could tell Kanga?

(A) 6
(B) 8
(C) 9
(D) 10
(E) 11

6 A bag contains balls of five different colors. Two are red, three are blue, ten are white, four are green, and three are black. Balls are taken from the bag without looking, and not returned. What is the smallest number of balls that should be taken from the bag to be sure that two balls of the same color have been taken?

(A) 2
(B) 12
(C) 10
(D) 5
(E) 6

7 Alex lights a candle every ten minutes. Each candle burns for 40 minutes and then goes out. How many candles are still burning 55 minutes after Alex lit the first candle?

(A) 2
(B) 3
(C) 4
(D) 5
(E) 6

8 The average number of children in five families cannot be

(A) 0.2
(B) 1.2
(C) 2.2
(D) 2.4
(E) 2.5

9 Mark and Liza stand on opposite sides of a circular fountain. They then start to run clockwise round the fountain. Mark's speed is $\frac{9}{8}$ of Liza's speed. How many circuits has Liza completed when Mark catches up with her for the first time?

(A) 4
(B) 8
(C) 9
(D) 2
(E) 72

10 The positive integers x, y, and z satisfy $x \cdot y = 14$, $y \cdot z = 10$, and $z \cdot x = 35$. What is the value of $x + y + z$?

(A) 10
(B) 12
(C) 14
(D) 16
(E) 18

4 Points Each

11 Carina and a friend are playing a game of "battleships" on a 5 × 5 board. Carina has already placed two ships as shown. She still has to place a 3 × 1 ship so that it covers exactly three cells. No two ships can have a point in common. How many positions are there for her 3 × 1 ship?

(A) 4
(B) 5
(C) 6
(D) 7
(E) 8

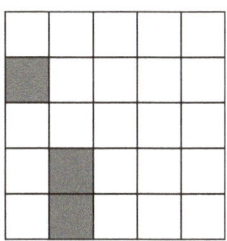

12 In the diagram, $\alpha = 55°$, $\beta = 40°$, and $y = 35°$. What is the value of δ?

(A) 100°
(B) 105°
(C) 120°
(D) 125°
(E) 130°

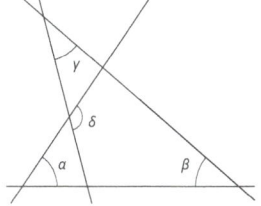

13 The perimeter of a trapezoid is 5 and the lengths of its sides are integers. What are the measures of the two smallest angles of the trapezoid?

(A) 30° and 30°
(B) 60° and 60°
(C) 45° and 45°
(D) 30° and 60°
(E) 45° and 90°

14 One of the following nets cannot be folded to form a cube. Which one?

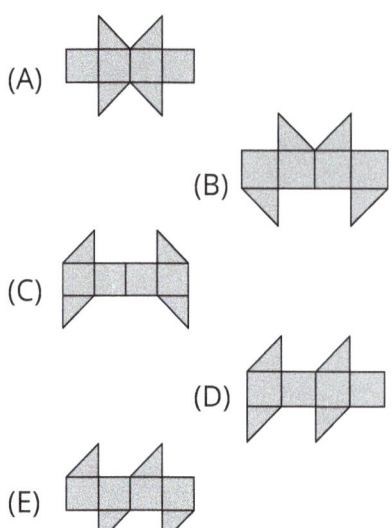

15 Vasya wrote down several consecutive integers. Which of the following could not be the percentage of odd numbers among them?

(A) 40
(B) 45
(C) 48
(D) 50
(E) 60

16 The edges of rectangle ABCD are parallel to the coordinate axes. ABCD lies below the x-axis and to the right of the y-axis, as shown in the figure. The coordinates of the four points A, B, C, and D are all integers. For each of these points we calculate the value y-coordinate ÷ x-coordinate. Which of the four points gives the least value?

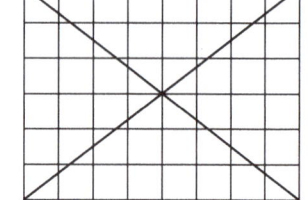

(A) A
(B) B
(C) C
(D) D
(E) It depends on the rectangle.

17 All 4-digit positive integers with the same four digits as in the number 2013 are written on the blackboard in an increasing order. What is the largest possible difference between two neighboring numbers on the blackboard?

(A) 702
(B) 703
(C) 693
(D) 793
(E) 198

18 In the 6 × 8 grid shown, 24 of the cells are not intersected by either diagonal. When the diagonals of a 6 × 10 grid are drawn, how many of the cells are not intersected by either diagonal?

(A) 28
(B) 29
(C) 30
(D) 31
(E) 32

19 Andy, Betty, Cathie, Dannie, and Eddy were born on 02/20/2001, 03/12/2000, 03/20/2001, 04/12/2000, and 04/23/2001 (month/day/year). Andy and Eddy were born in the same month. Also, Betty and Cathie were born in the same month. Andy and Cathie were born on the same day of different months. Also, Dannie and Eddy were born on the same day of different months. Which of these children is the youngest?

(A) Andy
(B) Betty
(C) Cathie
(D) Dannie
(E) Eddy

20 John made a building of cubes standing on a 4 × 4 grid. The diagram shows the number of cubes standing on each cell. When John looks from the back, what does he see?

Back

4	2	3	2
3	3	1	2
2	1	3	1
1	2	1	2

Front

(A)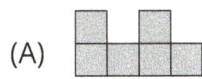

(B)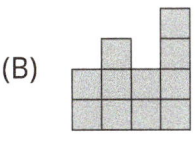

(C)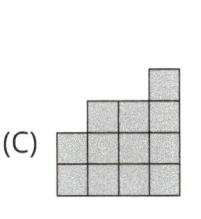

(D)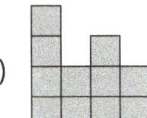

(E)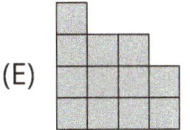

5 Points Each

21 The diagram shows a shaded quadrilateral *KLMN* drawn on a grid. Each cell of the grid has sides of length 2 cm. What is the area of *KLMN*?

(A) 96 cm²
(B) 84 cm²
(C) 76 cm²
(D) 88 cm²
(E) 104 cm²

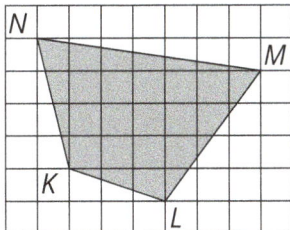

22 Let *S* be the number of squares among the integers from 1 to 2013^6. Let *Q* be the number of cubes among the same integers. Then

(A) $S = Q$
(B) $2S = 3Q$
(C) $3S = 2Q$
(D) $S = 2013Q$
(E) $S^3 = Q^2$

23 John chooses a 5-digit positive integer and deletes one of its digits to make a 4-digit number. The sum of this 4-digit number and the original 5-digit number is 52713. What is the sum of the digits of the original 5-digit number?

(A) 26
(B) 20
(C) 23
(D) 19
(E) 17

24 A gardener wants to plant twenty trees (maples and lindens) along an avenue in the park. The number of trees between any two maples must not be equal to three. Of these twenty trees, what is the greatest number of maples that the gardener can plant?

(A) 8
(B) 10
(C) 12
(D) 14
(E) 16

25. Andrew and Daniel recently took part in a marathon. After they had finished, they noticed that Andrew finished ahead of twice as many runners as finished ahead of Daniel, and that Daniel finished ahead of 1.5 times as many runners as finished ahead of Andrew. Andrew finished in 21st place. How many runners took part in the marathon?

(A) 31
(B) 41
(C) 51
(D) 61
(E) 81

26. Four cars enter a roundabout at the same time, each one from a different direction, as shown in the diagram. Each of the cars drives less than once around the roundabout, and no two cars leave the roundabout in the same direction. How many different ways are there for the cars to leave the roundabout?

(A) 9
(B) 12
(C) 15
(D) 24
(E) 81

27. A sequence starts with 1, −1, −1, 1, −1. After the fifth term, every term is equal to the product of the two preceding terms. For example, the sixth term is equal to the product of the fourth term and the fifth term. What is the sum of the first 2013 terms?

(A) −1006
(B) −671
(C) 0
(D) 671
(E) 1007

28. Ria bakes six raspberry pies one after the other, numbering them 1 to 6 in order, with the first being number 1. While she is doing this, her children sometimes run into the kitchen and eat the hottest pie. Which of the following could not be the order in which the pies are eaten?

(A) 123456
(B) 125436
(C) 325461
(D) 456231
(E) 654321

29. Each of the four vertices and six edges of a tetrahedron is marked with one of the ten numbers 1, 2, 3, 4, 5, 6, 7, 8, 9 and 11 (number 10 is omitted). Each number is used exactly once. For any two vertices of the tetrahedron, the sum of two numbers at these vertices is equal to the number on the edge connecting these two vertices. The edge PQ is marked with the number 9. Which number is used to mark edge RS?

(A) 4
(B) 5
(C) 6
(D) 8
(E) 11

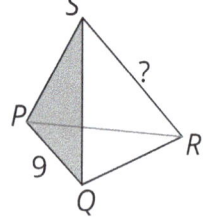

30. A positive integer N is smaller than the sum of its three greatest divisors (naturally, excluding N itself). Which of the following statements is true?

(A) All such integers N are divisible by 4.
(B) All such integers N are divisible by 5.
(C) All such integers N are divisible by 6.
(D) All such integers N are divisible by 7.
(E) There is no such integer N.

2015

2015

3 Points Each

1. $\dfrac{20}{15} =$

(A) $\dfrac{2+0+1+5}{1+5}$

(B) $\dfrac{2+0+1+5}{2+0}$

(C) $\dfrac{2+0}{1+5}$

(D) $\dfrac{20+15}{20}$

(E) $\dfrac{20+15}{15}$

---

2. A journey from Koice to Poprad through Preov lasts 2 hours and 10 minutes. The part of the journey from Koice to Preov lasts 35 minutes. How long does the part of the journey from Preov to Poprad last?

(A) 1 hour and 35 minutes
(B) 1 hour and 45 minutes
(C) 1 hour and 55 minutes
(D) 1 hour and 25 minutes
(E) 1 hour and 15 minutes

---

3. Four identical small rectangles are put together to form a large rectangle as shown. The length of the shorter side of the large rectangle is 10 cm. What is the length of the longer side of the large rectangle?

(A) 10 cm
(B) 20 cm
(C) 30 cm
(D) 40 cm
(E) 50 cm

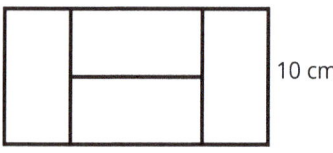

---

4. Which of the following numbers is closest to 2.015 × 510.2?

(A) 0.1
(B) 1
(C) 10
(D) 100
(E) 1000

---

5. The net of a cube with numbered faces is shown in the diagram. Sasha correctly adds the numbers on opposite faces of this cube. What three totals does Sasha get?

(A) 4, 6, 11
(B) 4, 5, 12
(C) 5, 6, 10
(D) 5, 7, 9
(E) 5, 8, 8

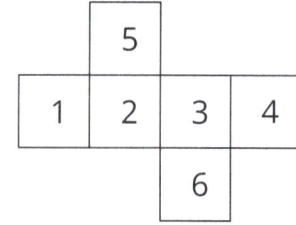

---

6. Which of the following numbers is not an integer?

(A) $\dfrac{2011}{1}$
(B) $\dfrac{2012}{2}$
(C) $\dfrac{2013}{3}$
(D) $\dfrac{2014}{4}$
(E) $\dfrac{2015}{5}$

7 A triangle has side lengths of 6, 10, and 11. An equilateral triangle has the same perimeter. What is the side length of the equilateral triangle?

(A) 18
(B) 11
(C) 10
(D) 9
(E) 6

8 The diagram shows the net of a triangular prism. Which edge coincides with edge UV when the net is folded to make the prism?

(A) WV
(B) XW
(C) XY
(D) QR
(E) RS

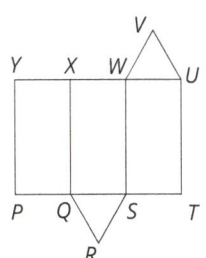

9 When Simon the squirrel comes down to the ground, he never goes further than 5 m from the trunk of his tree. However, he also stays at least 5 m away from the doghouse. Which of the following illustrations most accurately shows the shape of the region on the ground where Simon might go?

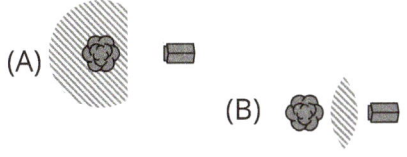

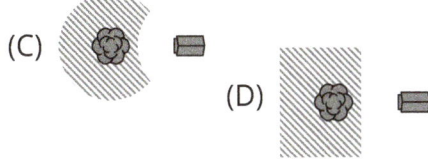

10 A cyclist rides at 5 m per second. The wheels of his bicycle have a circumference of 125 cm. How many complete rotations does each wheel make in 5 seconds?

(A) 4
(B) 5
(C) 10
(D) 20
(E) 25

4 Points Each

11 In a certain class, no two boys were born on the same day of the week and no two girls were born in the same month. If a new boy or a new girl joined this class, one of these two conditions would no longer be true. How many children are there in the class?

(A) 18
(B) 19
(C) 20
(D) 24
(E) 25

12 In the diagram, the center of the top square is directly above the common edge of the lower two squares. Each square has sides with a length of 1. What is the area of the shaded region?

(A) $\frac{3}{4}$
(B) $\frac{7}{8}$
(C) 1
(D) $1\frac{1}{4}$
(E) $1\frac{1}{2}$

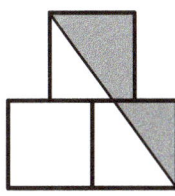

13 Every asterisk in the equation
$2*0*1*5*2*0*1*5*2*0*1*5=0$
is to be replaced with either + or – so that the equation is correct. What is the smallest number of asterisks that must be replaced with +?

(A) 1
(B) 2
(C) 3
(D) 4
(E) 5

---

14 During a rainstorm, 15 liters of water fell per square meter. By how much did the water level rise in an outdoor pool?

(A) 150 cm
(B) 0.15 cm
(C) 15 cm
(D) 1.5 cm
(E) It depends on the size of the pool.

---

15 A bush has 10 branches. Each branch has either 5 leaves only or 2 leaves and 1 flower. Which of the following could be the total number of leaves the bush has?

(A) 45
(B) 39
(C) 37
(D) 31
(E) None of (A) to (D).

---

16 The mean score of the students who took a mathematics test was 6. Exactly 60% of the students passed the test. The mean score of the students who passed the test was 8. What was the mean score of the students who failed the test?

(A) 1
(B) 2
(C) 3
(D) 4
(E) 5

---

17 One corner of a square is folded to the center of the square to form an irregular pentagon. The areas of the pentagon and of the square are consecutive integers. What is the area of the square?

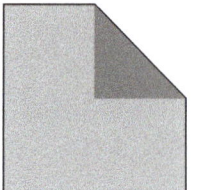

(A) 2
(B) 4
(C) 8
(D) 16
(E) 32

---

18 Rachel added the lengths of three sides of a rectangle and got 44 cm. Heather added the lengths of three sides of the same rectangle and got 40 cm. What is the perimeter of the rectangle?

(A) 42 cm
(B) 56 cm
(C) 64 cm
(D) 84 cm
(E) 112 cm

---

19 The diagram indicates the colors of some of the unit segments of a design (see the figure). Luis wants to color each remaining unit segment in the design one of the following colors: red, blue, or green. Each of the 6 triangles must have one side of each color. What color can he use for the segment marked x?

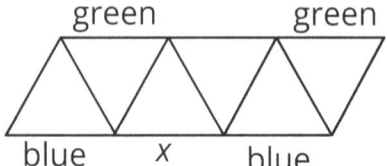

(A) only green
(B) only red
(C) only blue
(D) either red or blue
(E) This is impossible.

20 Irina asked five of her students how many of the five had studied the day before. Pol said none, Berta said only one, Ona said exactly two, Eugeni said exactly three, and Gerard said exactly four. Irina knew that those students who had not studied were not telling the truth, but those who had studied were telling the truth. How many of these students had studied the day before?

(A) 0
(B) 1
(C) 2
(D) 3
(E) 4

5 Points Each

21 Ria wants to write a number in each of the seven regions of the diagram below. Two regions are neighbors if they share part of their boundary. The number in each region needs to be the sum of the numbers in all its neighboring regions. Ria has already written in two of the numbers, as shown. What number does she need to write in the center region?

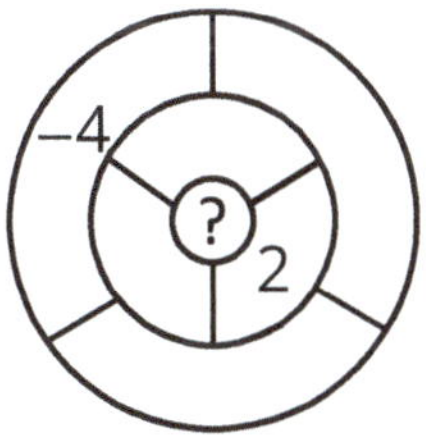

(A) 1
(B) −2
(C) 6
(D) −4
(E) 0

22 In a group of kangaroos, the two lightest kangaroos weigh 25% of the total weight of the group. The three heaviest kangaroos weigh 60% of the total weight. How many kangaroos are in the group?

(A) 6
(B) 7
(C) 8
(D) 15
(E) 20

23 In trapezoid ABCD, the sides AB and DC are parallel. The measure of angle CDA is 120° and $CD = DA = \frac{1}{3}AB$. What is the measure of angle ABC?

(A) 45°
(B) 30°
(C) 25°
(D) 22.5°
(E) 15°

24 Five positive integers, not necessarily all different, are written on five cards. Peter calculates the sum of the numbers on every pair of cards. He obtains only three different totals, 57, 70, and 83. What is the largest integer on any of the cards?

(A) 35
(B) 42
(C) 48
(D) 53
(E) 82

25 Mary wrote down the remainders obtained by dividing the number 2015 by each of the numbers 1, 2, 3, and so on, up to and including 1000. What is the largest of these remainders?

(A) 503
(B) 504
(C) 671
(D) 672
(E) Some other value.

26 A square with an area of 30 is divided in two by a diagonal and then into triangles, as shown below. The areas of some of these triangles are given in the diagram. Which part of the diagonal is the longest?

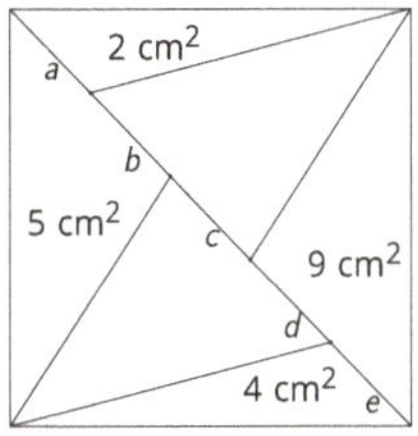

(A) a
(B) b
(C) c
(D) d
(E) e

27 Every positive integer is to be colored according to the following two rules.

(i) Each number is either red or green.
(ii) The sum of any two different numbers of a given color is that same color.

In how many different ways can this be done?

(A) 0
(B) 2
(C) 4
(D) 6
(E) more than 6

28 Five points lie on a line. Alex finds the distances between every possible pair of points. He obtains, in increasing order, 2, 5, 6, 8, 9, k, 15, 17, 20, and 22. What is the value of k?

(A) 10
(B) 11
(C) 12
(D) 13
(E) 14

29 The figure shows a sheet with points marked. The distance from one point to the next is the same both horizontally and vertically. Let four points at a time become vertices of different squares. How many squares with different areas is it possible to make?

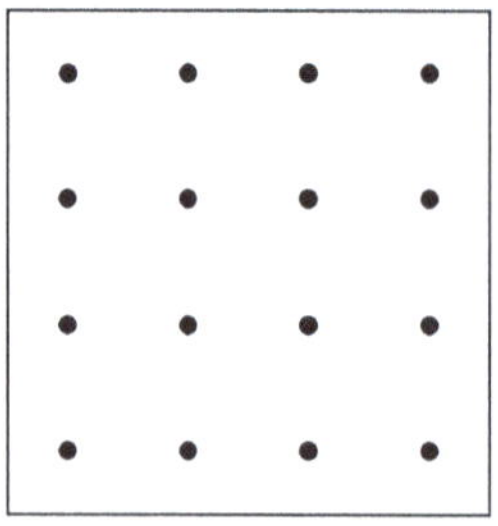

(A) 2
(B) 3
(C) 4
(D) 5
(E) 6

30 We need to write each of the numbers from 1 to 9 in the boxes of the diagram shown below in such a way that the result of each operation indicated is in the box to which the arrow points. Which number do we need to write in the box marked with the question mark?

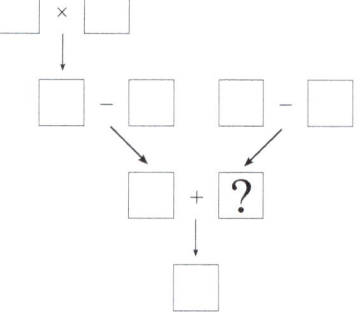

(A) 2
(B) 3
(C) 5
(D) 6
(E) 7

2017

2017

3 Points Each

1 What is the time 17 hours after 5:00 p.m.?

(A) 8:00 a.m.
(B) 10:00 a.m.
(C) 11:00 a.m.
(D) 12:00 p.m.
(E) 1:00 p.m.

2 A group of girls stands in a circle. Xena is the fourth to the left from Yana and the seventh to the right from Yana. How many girls are in the group?

(A) 9
(B) 10
(C) 11
(D) 12
(E) 13

3 What number must be subtracted from −17 to obtain −33?

(A) −50
(B) −16
(C) 16
(D) 40
(E) 50

4 The diagram shows a striped isosceles triangle and its altitude. Each stripe has the same height. What fraction of the area of the triangle is white?

(A) $\frac{1}{2}$
(B) $\frac{1}{3}$
(C) $\frac{2}{3}$
(D) $\frac{3}{4}$
(E) $\frac{2}{5}$

5 Which of the following equalities is correct?

(A) $\frac{4}{1} = 1.4$
(B) $\frac{5}{2} = 2.5$
(C) $\frac{6}{3} = 3.6$
(D) $\frac{7}{4} = 4.7$
(E) $\frac{8}{5} = 5.8$

6 The diagram shows two rectangles whose sides are parallel. What is the difference in the lengths of the perimeters of the two rectangles?

(A) 12 m
(B) 16 m
(C) 20 m
(D) 21 m
(E) 24 m

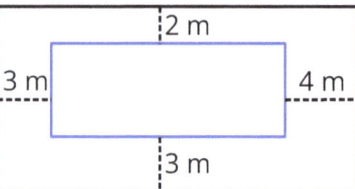

7 Bob folded a piece of paper twice and then cut one hole in the folded piece of paper. When he unfolded the paper, he saw the arrangement shown in the diagram.

How had Bob folded his piece of paper?

(A)

(B)

(C)

(D)

(E)

8. The sum of three different positive integers is 7. What is the product of these three integers?

(A) 12
(B) 10
(C) 9
(D) 8
(E) 5

9. The diagram shows four overlapping hearts. The areas of the hearts are 1 cm², 4 cm², 9 cm², and 16 cm². What is the shaded area?

(A) 9 cm²
(B) 10 cm²
(C) 11 cm²
(D) 12 cm²
(E) 13 cm²

10. Yvonne has 20 euros. Each of her four sisters has 10 euros. How many euros does Yvonne have to give to each of her sisters so that each of the five girls has the same amount of money?

(A) 2
(B) 4
(C) 5
(D) 8
(E) 10

4 Points Each

11. Annie the Ant started at the left end of a pole and crawled $\frac{2}{3}$ of its length. Bob the Beetle started at the right end of the same pole and crawled $\frac{3}{4}$ of its length. What fraction of the length of the pole are Annie and Bob now apart?

(A) $\frac{3}{8}$
(B) $\frac{1}{12}$
(C) $\frac{5}{7}$
(D) $\frac{1}{2}$
(E) $\frac{5}{12}$

12. One sixth of the audience in a children's theater were adults. Two fifths of children were boys. What fraction of the audience were girls?

(A) $\frac{1}{2}$
(B) $\frac{1}{3}$
(C) $\frac{1}{4}$
(D) $\frac{1}{5}$
(E) $\frac{2}{5}$

13. In the diagram, the dashed line and the black path form seven equilateral triangles. The length of the dashed line is 20. What is the length of the black path?

(A) 25
(B) 30
(C) 35
(D) 40
(E) 45

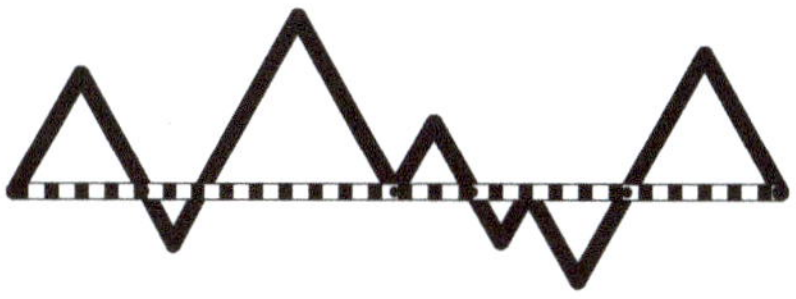

14 Four cousins, Ema, Iva, Rita, and Zina, are 3, 8, 12 and 14 years old, although not necessarily in that order. Ema is younger than Rita. The sum of the ages of Zina and Ema is divisible by 5. The sum of the ages of Zina and Rita is also divisible by 5. How old is Iva?

(A) 14
(B) 12
(C) 8
(D) 5
(E) 3

15 This year there were more than 800 runners participating in the Kangaroo Hop. Exactly 35% of the runners were women and there were 252 more men than women. How many runners were there in total?

(A) 802
(B) 810
(C) 822
(D) 824
(E) 840

16 Ria wants to write a number in each box of the diagram shown. She has already written two of the numbers. She wants the sum of all the numbers to equal 35, the sum of the numbers in the first three boxes to equal 22, and the sum of the numbers in the last three boxes to equal 25. What is the product of the numbers she writes in the gray boxes?

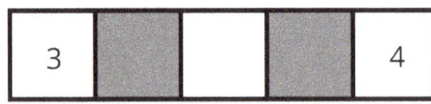

(A) 63
(B) 108
(C) 0
(D) 48
(E) 39

17 Simon wants to cut a piece of thread into nine pieces of the same length and marks his cutting points. Barbara wants to cut the same piece of thread into only eight pieces of the same length and also marks her cutting points. Carl then cuts the thread at all the cutting points that are marked. How many pieces of thread does Carl obtain?

(A) 15
(B) 16
(C) 17
(D) 18
(E) 19

18 Two segments, each 1 cm long, are marked on opposite sides of a square with a side length of 8 cm. The ends of the segments are joined as shown in the diagram. What is the shaded area, in cm²?

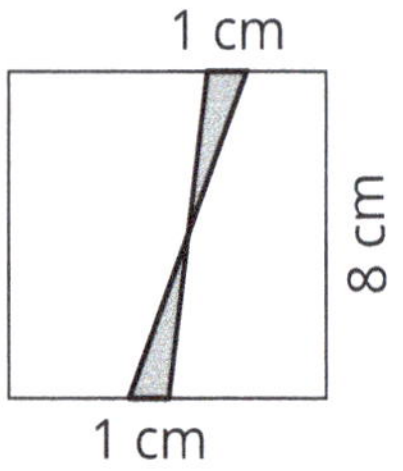

(A) 2
(B) 4
(C) 6.4
(D) 8
(E) 10

19 Tycho wants to prepare a schedule for his jogging. He wants to jog exactly twice a week, and on the same days every week. He never wants to jog on two consecutive days. How many different schedules can he prepare?

(A) 16
(B) 14
(C) 12
(D) 10
(E) 8

20 Emily wants to write a number into each cell of a 3 × 3 table so that the sum of the numbers in any two cells that share an edge are the same. She has already written two numbers, as shown in the diagram. What is the sum of all the numbers in the table?

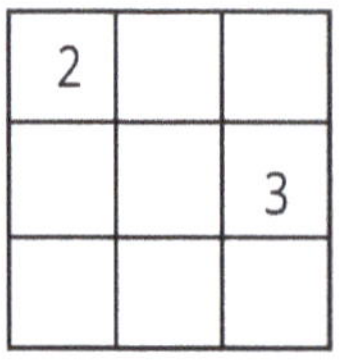

(A) 18
(B) 20
(C) 21
(D) 22
(E) 23

5 Points Each

21 The numbers of degrees of the angles in a triangle are three different integers. What is the minimum possible sum of its smallest and largest angles?

(A) 61°
(B) 90°
(C) 91°
(D) 120°
(E) 121°

22 Ten kangaroos stood in a line as shown in the diagram. At some point, two kangaroos standing side by side and facing each other exchanged places by jumping past each other. This was repeated until no further jumps were possible. How many exchanges were made?

(A) 15
(B) 16
(C) 18
(D) 20
(E) 21

23 Diana has nine numbers: 1, 2, 3, 4, 5, 6, 7, 8, and 9. She adds 2 to some of them, and 5 to all the others. What is the smallest number of different results she can obtain?

(A) 5
(B) 6
(C) 7
(D) 8
(E) 9

24 Buses leave the airport every 3 minutes to drive to the city center. A car leaves the airport at the same time as one of the buses and drives to the city center by the same route. It takes each bus 60 minutes and the car 35 minutes to drive from the airport to the city center. How many buses does the car pass on its way to the center, not including the bus it left with?

(A) 8
(B) 9
(C) 10
(D) 11
(E) 13

25 Olesia's tablecloth has a regular pattern, as shown in the diagram. What percentage of the tablecloth is black?

(A) 16
(B) 24
(C) 25
(D) 32
(E) 36

26 Each digit in the sequence starting with 2, 3, 6, 8, 8 is obtained in the following way: the first two digits are 2 and 3 and afterwards each digit is the last digit of the product of the two preceding digits in the sequence. What is the 2017th digit in the sequence?

(A) 2
(B) 3
(C) 4
(D) 6
(E) 8

27 Mike had 125 small cubes. He glued some of them together to form a big cube with nine tunnels leading through the whole cube as shown in the diagram. How many of the small cubes did he not use?

(A) 52
(B) 45
(C) 42
(D) 39
(E) 36

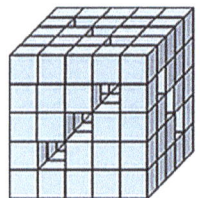

28 Two runners are training on a 720-meter circular track. They run in opposite directions, each at a constant speed. The first runner takes four minutes to complete a full lap and the second runner takes five minutes. How many meters does the second runner run between two consecutive meetings of the two runners?

(A) 355
(B) 350
(C) 340
(D) 330
(E) 320

29 Sarah wants to write a positive integer in each box in the diagram so that each number above the bottom row is the sum of the two numbers in the boxes immediately underneath. What is the largest number of odd numbers that Sarah can write?

(A) 5
(B) 7
(C) 8
(D) 10
(E) 11

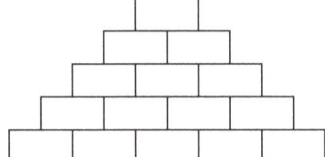

30 The diagram shows parallelogram ABCD with area S. The intersection point of the diagonals of the parallelogram is O. Point M is marked on DC. The intersection point of AM and BD is E and the intersection point of BM and AC is F. The sum of the areas of the triangles AED and BFC is $\frac{1}{3}S$. What is the area of the quadrilateral EOFM, in terms of S?

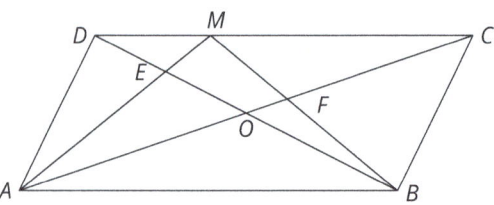

(A) $\frac{1}{6}S$
(B) $\frac{1}{8}S$
(C) $\frac{1}{10}S$
(D) $\frac{1}{12}S$
(E) $\frac{1}{14}S$

2019

2019

3 Points Each

1 Which cloud contains four even numbers?

(A) 3 9 6 5

(B) 33 3 13 23

(C) 3 30 27 9

(D) 1 9 6 3

(E) 10 2 34 58

2 How many hours are there in ten quarters of an hour?

(A) 40
(B) 5 and a half
(C) 4
(D) 3
(E) 2 and a half

3 A 3 × 3 × 3 cube is built from 1 × 1 × 1 cubes. Then some cubes are removed from front to back, from left to right, and from top to bottom, as shown. How many 1 × 1 × 1 cubes are left?

(A) 15
(B) 18
(C) 20
(D) 21
(E) 22

4 Three rings are linked as shown in the diagram. Which of the following diagrams also shows the three rings linked in the same way?

(A)  (B)

(C) 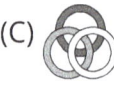 (D)

(E)

5 Which of the diagrams below cannot be drawn without lifting your pencil off the page or drawing along the same line twice?

(A)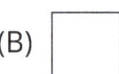

(B) □

(C) ◨

(D)

(E) ⬒

6 Five friends met. Each of them gave a cupcake to each of the others. They then ate all the cupcakes they had been given. As a result, the total number of cupcakes they had decreased by half. How many cupcakes did the five friends have at the start?

(A) 20
(B) 24
(C) 30
(D) 40
(E) 60

7 In a race, Lotar finished before Manfred, Victor finished after Jan, Manfred finished before Jan, and Eddy finished before Victor. Who finished last of these five runners?

(A) Victor
(B) Manfred
(C) Lotar
(D) Jan
(E) Eddy

8 The pages of the book Juliet is reading are all numbered. The numbers used on the pages contain the digit 0 exactly five times and the digit 8 exactly six times. What is the number on the final page?

(A) 48
(B) 58
(C) 60
(D) 68
(E) 88

9 A large square is divided into smaller squares. What fraction of the large square is colored gray?

(A) $\frac{2}{3}$
(B) $\frac{2}{5}$
(C) $\frac{4}{7}$
(D) $\frac{4}{9}$
(E) $\frac{5}{12}$

---

10 Andrew divided some apples into six equal piles. Boris divided the same number of apples into five equal piles. Boris noticed that each of his piles contains two more apples than each of Andrew's piles. How many apples does Andrew have?

(A) 60
(B) 65
(C) 70
(D) 75
(E) 80

---

4 Points Each

11 A different four-digit number was written on each of three pieces of paper. The pieces of paper are arranged so that three of the digits are covered, as shown. The sum of the three four-digit numbers is 10126. Which are the covered digits?

(A) 5, 6, and 7
(B) 4, 5, and 7
(C) 4, 6, and 7
(D) 4, 5, and 6
(E) 3, 5, and 6

---

12 In the diagram, $PQ = PR = QS$, and angle $\angle QPR = 20°$. What is the size of angle $\angle RQS$?

(A) 50°
(B) 60°
(C) 65°
(D) 70°
(E) 75°

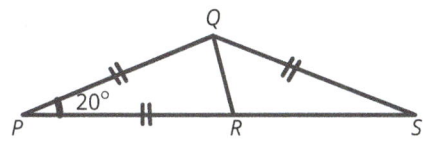

---

13 Which of the following 4 × 4 tiles cannot be formed by combining the two given pieces?

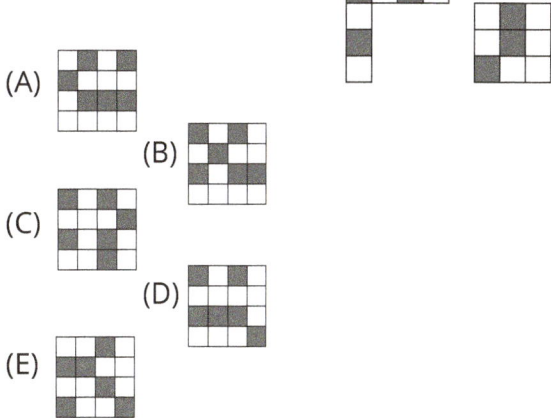

(A)
(B)
(C)
(D)
(E)

---

14 Alan, Bella, Claire, Dora, and Erik met at a party and shook hands exactly once with everyone they already knew. Alan shook hands once, Bella shook hands twice, Claire shook hands three times, and Dora shook hands four times. How many times did Erik shake hands?

(A) 1
(B) 2
(C) 3
(D) 4
(E) 0

---

15 Jane is playing basketball. In a series of 20 shots, Jane scored 55% of the time. After five more shots, her scoring rate increased to 56%. On how many of the last five shots did she score?

(A) 1
(B) 2
(C) 3
(D) 4
(E) 5

16 Cathie folded a square sheet of paper exactly in half twice and then cut it in the middle twice, as shown in the diagram. How many of the pieces that she obtains are squares?

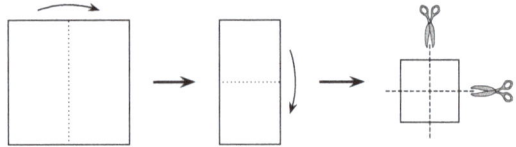

(A) 3
(B) 4
(C) 5
(D) 6
(E) 8

17 Michael has dogs, cows, cats, and kangaroos as pets. He tells Helen that he has 24 pets in total and that $\frac{1}{8}$ of them are dogs, $\frac{3}{4}$ are NOT cows, and $\frac{2}{3}$ are NOT cats. How many kangaroos does Michael have?

(A) 4
(B) 5
(C) 6
(D) 7
(E) 8

18 Some identical rectangles are drawn on the floor. A triangle with a base of 10 cm and a height of 6 cm is drawn over them, as shown, and the region inside the rectangles and outside the triangle is shaded. What is the area of the shaded region?

(A) 10 cm²
(B) 12 cm²
(C) 14 cm²
(D) 15 cm²
(E) 21 cm²

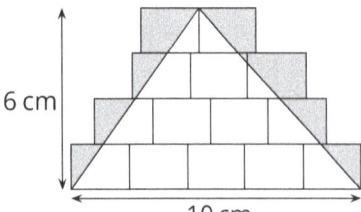

19 Julio had two cylindrical candles with different heights and diameters. The first candle lasts 6 hours, while the second candle lasts 8 hours. He lit both candles at the same time and three hours later both candles were the same height. What was the ratio of their original heights?

(A) 4:3
(B) 8:5
(C) 5:4
(D) 3:5
(E) 7:3

20 Aylin wants to create a path of matches using as few matches as possible. She places each match on the piece of paper like the one shown, along one of the dotted lines. Her path returns to the left-hand end of her original match. The numbers shown in some of the cells are equal to the number of matches around that cell. How many matches are in this path?

(A) 12
(B) 14
(C) 16
(D) 18
(E) 20

5 Points Each

21 The integers from 1 to n, inclusive, are equally spaced in order around a circle. The diameter through the position of the integer 7 also goes through the position of 23, as shown. What is the value of n?

(A) 30
(B) 32
(C) 34
(D) 36
(E) 38

22 Liam spent all his money buying 50 soda bottles at the store for $1 each. Hoping to make a profit, he raises the price and sells all 50 soda bottles at the new higher price. After selling 40 bottles, he has $10 more than he started with. He then sells all the remaining bottles. How much money does Liam now have?

(A) $70
(B) $75
(C) $80
(D) $90
(E) $100

23 Natasha has many sticks of length 1. Each stick is colored blue, red, yellow, or green. She wants to make a 3 × 3 grid, as shown, so that each 1 × 1 square in the grid has four sides of different colors. What is the smallest number of green sticks that she could use?

(A) 3
(B) 4
(C) 5
(D) 6
(E) 7

24 An ant would like to walk along a marked line on the surface of a cube until it returns to its starting point. From which one of the following nets could a cube be made so that such a journey is possible?

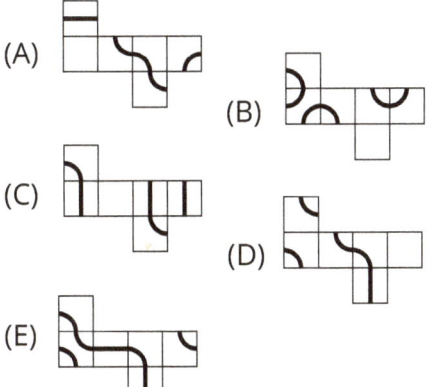

25 Elisabeta had a large bag of 60 chocolates. She started by eating one-tenth of them on Monday, then one-ninth of the remainder on Tuesday, then one-eighth of the rest on Wednesday, then one-seventh on Thursday, and so on until she eats half of the chocolates remaining from the previous day. How many chocolates does she have left?

(A) 1
(B) 2
(C) 3
(D) 4
(E) 6

26 Prab painted each of the eight circles in the diagram red, yellow, or blue in such a way that no two circles that are joined directly are painted the same color. Which two circles are necessarily painted the same color?

(A) 5 and 8
(B) 1 and 6
(C) 2 and 7
(D) 4 and 5
(E) 3 and 6

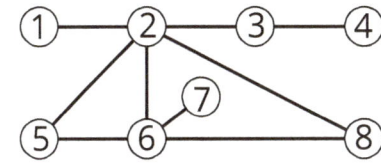

27 When Ria and Flora compared their savings, they found that the ratio of their savings was 5:3. Then Ria bought a tablet for $160 and the ratio of their savings changed to 3:5. How many dollars did Ria have before buying the tablet?

(A) 192
(B) 200
(C) 250
(D) 400
(E) 420

28 Several three-player teams enter a chess tournament. Each player in a team plays exactly once against every player from all the other teams. For organizational reasons, no more than 250 games can be played in total. At most how many teams can enter the tournament?

(A) 11
(B) 10
(C) 9
(D) 8
(E) 7

29 The diagram shows the square ABCD with P, Q, and R as the midpoints of the sides DA, BC, and CD respectively. What fraction of the square ABCD is shaded?

(A) $\frac{3}{4}$
(B) $\frac{5}{8}$
(C) $\frac{1}{2}$
(D) $\frac{7}{16}$
(E) $\frac{3}{8}$

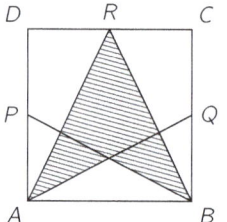

30 A train is made up of 18 cars. There are 700 passengers traveling on the train. In any block of five adjacent cars, there are 199 passengers in total. How many passengers are in the middle two cars of the train?

(A) 70
(B) 77
(C) 78
(D) 96
(E) 103

2021

3 Points Each

1 Which of the following symbols for signs of the zodiac has an axis of symmetry?

(A) Sagittarius

(B) Scorpio

(C) Leo

(D) Cancer

(E) Capricorn

2 The figure shows three concentric circles with four lines passing through their common center. What percentage of the figure is shaded?

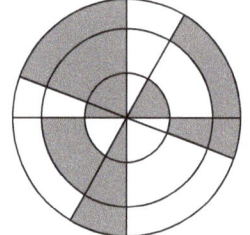

(A) 30%
(B) 35%
(C) 40%
(D) 45%
(E) 50%

3 What is the value of $\dfrac{20 \cdot 21}{2+0+2+1}$?

(A) 42
(B) 64
(C) 80
(D) 84
(E) 105

4 How many four-digit numbers have the property that their digits, from left to right, are consecutive and in ascending order?

(A) 5
(B) 6
(C) 7
(D) 8
(E) 9

5 When the five pieces shown fit together correctly, the result is a rectangle with a calculation written on it. What is the result of this calculation?

(A) −100
(B) −8
(C) −1
(D) 199
(E) 208

6 Each of the five vases shown has the same height and each has a volume of 1 liter. Half a liter of water is poured into each vase. In which vase is the level of the water the highest?

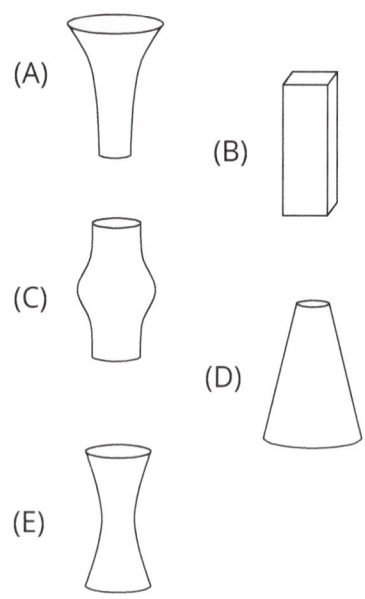

7 A student correctly added the two two-digit numbers on the left of the board and got the answer 137. What answer will he get if he adds the two four-digit numbers on the right of the board?

(A) 13737
(B) 13837
(C) 14747
(D) 23737
(E) 137137

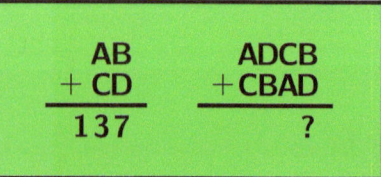

8. A 3 × 3 × 3 cube is made from white, gray, and black 1 × 1 × 1 cubes, as shown in the first diagram. The other two diagrams show the white part and the black part of the cube. Which of the following diagrams shows the gray part?

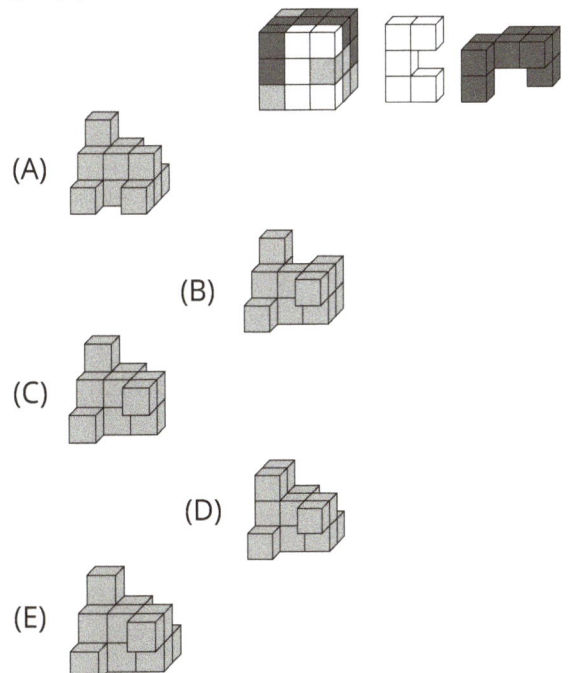

9. A bike lock has four wheels numbered with the digits 0 to 9 in order. Each of the four wheels is rotated by 180° from the code shown in the first diagram to get the correct code. What is the correct code for the bike lock?

(A)

(B)

(C)

(D)

(E) 8436

10. Byron is 5 cm taller than Aaron, but 10 cm shorter than Caron. Darren is 10 cm taller than Caron, but 5 cm shorter than Erin. Which of the following statements is true?

(A) Aaron and Erin are equal in height.
(B) Aaron is 10 cm taller than Erin.
(C) Aaron is 10 cm shorter than Erin.
(D) Aaron is 30 cm taller than Erin.
(E) Aaron is 30 cm shorter than Erin.

4 Points Each

11. A rectangular chocolate bar consists of a grid of congruent chocolate squares. The bar can be broken into two pieces by snapping it along a grid line. Neil breaks off a strip of chocolate that is two squares wide and eats the 12 squares he obtains. Later, from the same bar, Jack breaks off a strip of chocolate that is one square wide and eats the 9 squares he obtains. How many squares of chocolate are left in the bar?

(A) 72
(B) 63
(C) 54
(D) 45
(E) 36

12. A jar one-fifth filled with water weighs 560 g. The same jar four-fifths filled with water weighs 740 g. What is the weight of the empty jar?

(A) 60 g
(B) 112 g
(C) 180 g
(D) 300 g
(E) 500 g

13 The area of the large square is 16 cm² and the area of each small square is 1 cm². What is the total area of the black flower?

(A) 3 cm²
(B) $\frac{7}{2}$ cm²
(C) 4 cm²
(D) $\frac{11}{2}$ cm²
(E) 6 cm²

14 Costa is building a new fence in his garden. He uses 25 planks of wood, each of which is 30 cm long. He arranges these planks so that each pair of two adjacent planks overlaps by the same amount.

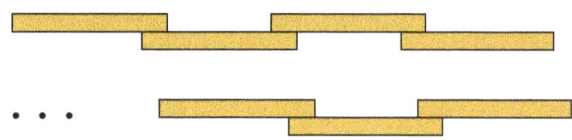

The total length of Costa's new fence is 6.9 meters. What is the length of the overlap within each pair of adjacent planks, in centimeters?

(A) 2.4
(B) 2.5
(C) 3
(D) 4.8
(E) 5

15 Five identical right-angled triangles can be arranged so that their larger acute angles touch to form the star shown in the diagram. It is also possible to form a different star by arranging more of these triangles so that their smaller acute angles touch. How many triangles are needed to form the second star?

(A) 10
(B) 12
(C) 18
(D) 20
(E) 24

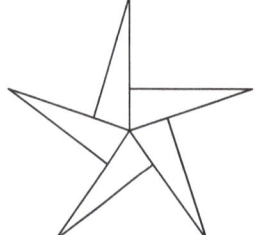

16 Five squares are positioned as shown. The small square indicated has an area of 1, as shown. What is the value of h?

(A) 3
(B) 3.5
(C) 4
(D) 4.2
(E) 4.5

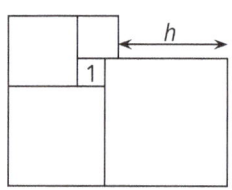

17 There are 20 questions in a quiz. Each participant scores 7 points for each correct answer, loses 4 points for each incorrect answer, and scores 0 points for each question left blank. Eric took the quiz and scored 100 points. How many questions did he leave blank?

(A) 0
(B) 1
(C) 2
(D) 3
(E) 4

18 A rectangular strip of paper with dimensions 4 × 13 is folded as shown in the diagram. Two rectangles are formed with areas P and Q where $P = 2Q$. What is the value of x?

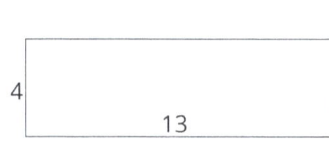
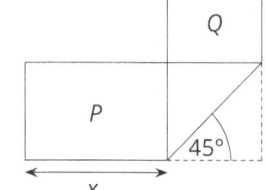

(A) 5
(B) 5.5
(C) 6
(D) 6.5
(E) $4\sqrt{2}$

19 A box of fruit contains twice as many apples as pears. Christy and Lily divided them up so that Christy had twice as many pieces of fruit as Lily. Which one of the following statements is always true?

(A) Christy took at least one pear.
(B) Christy took twice as many apples as pears.
(C) Christy took twice as many apples as Lily.
(D) Christy took as many apples as Lily got pears.
(E) Christy took as many pears as Lily got apples.

20 Three villages are connected by paths as shown. From Downend to Uphill, the detour via Middleton is 1 km longer than the direct wiggly path shown. From Downend to Middleton, the detour via Uphill is 5 km longer than the direct wiggly path. From Uphill to Middleton, the detour via Downend is 7 km longer than the direct wiggly path. How long is the shortest of the three direct wiggly paths between the villages?

(A) 1 km
(B) 2 km
(C) 3 km
(D) 4 km
(E) 5 km

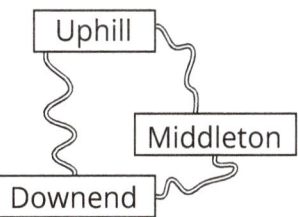

5 Points Each

21 In a particular fraction the numerator and denominator are both positive. The numerator of this fraction is increased by 40%. By what percentage should its denominator be decreased so that the new fraction is double the original fraction?

(A) 10%
(B) 20%
(C) 30%
(D) 40%
(E) 50%

22 A triangular pyramid is built with 20 cannon balls, as shown. Each cannon ball is labeled with one of the letters A, B, C, D, or E. There are four cannon balls with each type of label. The illustration shows the labels on the cannon balls on three of the faces of the pyramid. What is the label on the hidden cannon ball in the middle of the fourth face?

(A) A
(B) B
(C) C
(D) D
(E) E

23 The 6-digit number $\overline{2ABCDE}$ is multiplied by 3 and the result is the 6-digit number $\overline{ABCDE2}$. What is the sum of the digits of this number?

(A) 24
(B) 27
(C) 30
(D) 33
(E) 36

24 A box contains only green, red, blue, and yellow counters. There is always at least one green counter among any 27 counters chosen from the box; always at least one red among any 25 counters chosen; always at least one blue among any 22 counters chosen; and always at least one yellow among any 17 counters chosen. What is the largest number of counters that could be in the box?

(A) 27
(B) 29
(C) 51
(D) 87
(E) 91

25. A soccer ball is made of white hexagons and black pentagons, as seen in the illustration. There are a total of 12 pentagons. How many hexagons are there?

(A) 12
(B) 15
(C) 18
(D) 20
(E) 24

26. 2021 colored kangaroos are arranged in a row and are numbered from 1 to 2021. Each kangaroo is colored red, gray, or blue. Among any three consecutive kangaroos, there are always kangaroos of all three colors. Bruce guesses the colors of five kangaroos. These are his guesses: Kangaroo 2 is gray; Kangaroo 20 is blue; Kangaroo 202 is red; Kangaroo 1002 is blue; Kangaroo 2021 is gray. Only one of his guesses is wrong. What is the number of the kangaroo whose color he guessed incorrectly?

(A) 2
(B) 20
(C) 202
(D) 1002
(E) 2021

27. A 3 × 4 × 5 cuboid consists of 60 identical small cubes. A termite eats its way along the diagonal from P to Q. This diagonal does not intersect the edges of any small cube inside the cuboid. How many of the small cubes does it pass through on its journey?

(A) 8
(B) 9
(C) 10
(D) 11
(E) 12

28. In a town there are 21 knights who always tell the truth and 2000 knaves who always lie. A wizard divided 2020 of these 2021 people into 1010 pairs. Every person in a pair described the other person as either a knight or a knave. As a result, 2000 people were called knights and 20 people were called knaves. How many pairs of two knaves were there?

(A) 980
(B) 985
(C) 990
(D) 995
(E) 1000

29. In a tournament each of the six teams plays one match against each of the other teams. In each round, three matches take place simultaneously. A TV station has already decided which match it will broadcast for each round, as shown in the diagram. In which round will team D play against team F?

(A) 1
(B) 2
(C) 3
(D) 4
(E) 5

1	2	3	4	5
A – B	C – D	A – E	E – F	A – C

30. The diagram shows a quadrilateral divided into four smaller quadrilaterals with a common vertex K. The other labeled points divide the sides of the large quadrilateral into three equal parts. The numbers indicate the areas of the corresponding small quadrilaterals. What is the area of the shaded quadrilateral?

(A) 4
(B) 5
(C) 6
(D) 6.5
(E) 7

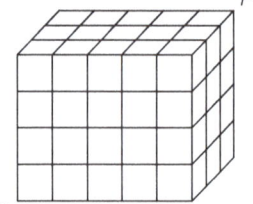

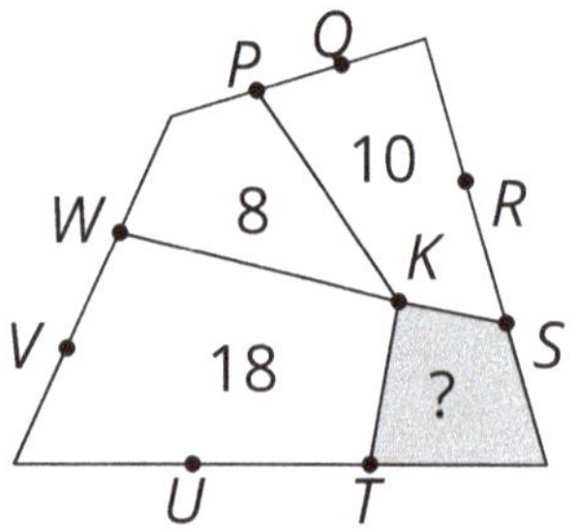

2023

2023

3 Points Each

1 The diagram shows a set of horizontal and vertical lines with one part removed. Which of the following could be the missing part?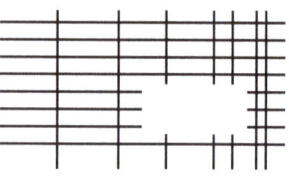

(A)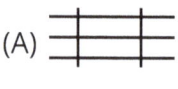

(B)

(C)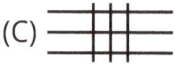

(D)

(E)

2 Which of the shapes below cannot be divided into two trapezoids by drawing a single straight line?

(A) triangle

(B) rectangle

(C) trapezoid

(D) regular hexagon

(E) square

3 A gray circle with two holes in it is placed on top of a clock face, as shown. The gray circle is turned around its center so that an 8 appears in one hole. Which two numbers could be seen in the other hole?

(A) 4 or 12
(B) 1 or 5
(C) 1 or 4
(D) 7 or 11
(E) 5 or 12

 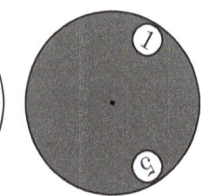

4 Pranav wants to write a number at each vertex and on each edge of the rhombus shown. He wants the sum of the numbers at the two vertices at the ends of each edge to be equal to the number written on the edge. What number will he write instead of the question mark?

(A) 11
(B) 12
(C) 13
(D) 14
(E) 15

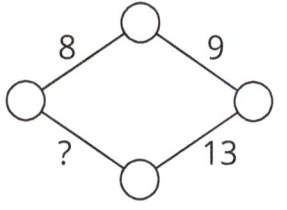

5 Kristina has a piece of transparent paper with some lines marked on it. She folds it along the dashed line. What can she now see?

(A)

(B)

(C)

(D)

(E)

6 A carpenter wants to tile a floor of dimensions 4 × 6 using identical tiles. No overlaps or gaps are allowed. Which of the following tiles could not be used?

(A)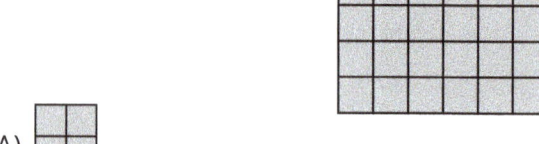

(B)

(C)

(D)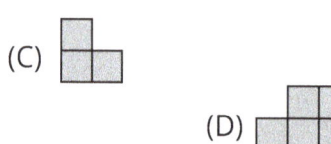

(E)

7 John has 150 coins. When he throws them on the table, 40% of them show heads and 60% of them show tails. How many coins showing tails does he need to turn over to have the same number of coins show heads as tails?

(A) 10
(B) 15
(C) 20
(D) 25
(E) 30

8 The diagram shows the initial position, the direction of travel, and how far four bumper cars move in five seconds. Which two cars will collide?

(A) A and B
(B) A and C
(C) A and D
(D) B and C
(E) C and D

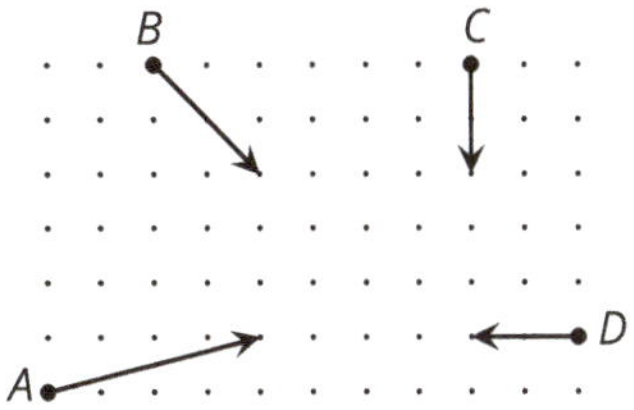

9 Anna has five circular discs, each of a different size. She decides to build a tower using three of her discs so that each disc in her tower is smaller than the disc below it. How many different towers could Anna construct?

(A) 5
(B) 6
(C) 8
(D) 10
(E) 15

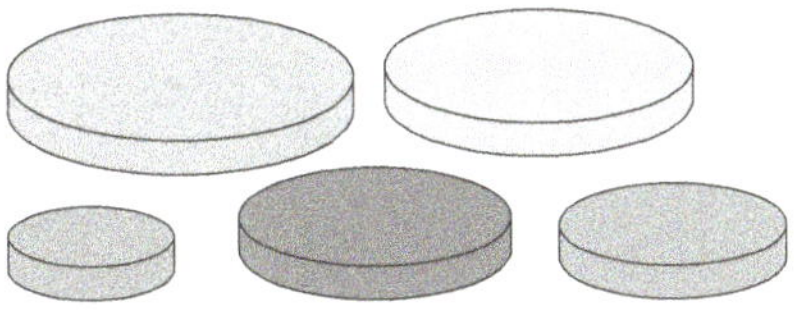

10 Evita wants to write the numbers 1 to 8 in the boxes of the grid shown so that the sums of the numbers in the boxes in each row are equal and the sums of the numbers in the boxes in each column are equal. She has already written numbers 3, 4, and 8 as shown. What number will she write in the shaded box?

(A) 1
(B) 2
(C) 5
(D) 6
(E) 7

4 Points Each

11 Anika wrote down three consecutive whole numbers in increasing order, but instead of digits she used symbols and wrote □◇◇, ♡△△, ♡△□. What would she write next?

(A) ♡♡◇
(B) □♡□
(C) ♡△◇
(D) ♡◇□
(E) ♡△♡

12 The diagram shows five equal semicircles and the lengths of some line segments. What is the radius of the semicircles?

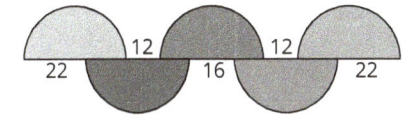

(A) 12
(B) 16
(C) 18
(D) 22
(E) 36

13 Some edges of a cube are to be colored red so that every face of the cube has at least one red edge. What is the smallest possible number of edges that could be colored red?

(A) 2
(B) 3
(C) 4
(D) 5
(E) 6

14 Matchsticks can be used to write digits, as shown in the diagram.

How many different positive integers can be written using exactly six matchsticks in this way?

(A) 2
(B) 4
(C) 6
(D) 8
(E) 9

15 The edges of a square are 1 cm long. How many points on the plane are exactly 1 cm away from two of the vertices of this square?

(A) 4
(B) 6
(C) 8
(D) 10
(E) 12

---

16 Triangle *ABC* is isosceles with ∠*ABC* = 40°. The two marked angles, ∠*EAB* and ∠*DCA*, are equal. What is the size of the angle ∠*CFE*?

(A) 55°
(B) 60°
(C) 65°
(D) 70°
(E) 75°

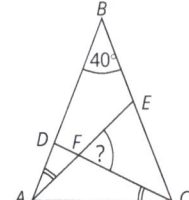

---

17 Tom, John, and Lily each shot six arrows at a target. Arrows hitting anywhere within the same ring score the same number of points. Tom scored 46 points and John scored 34 points, as shown. How many points did Lily score?

(A) 37
(B) 38
(C) 39
(D) 40
(E) 41

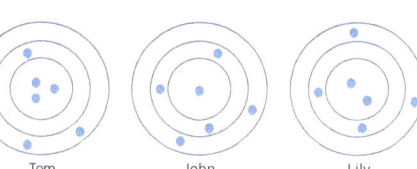

---

18 The diagram shows a rectangle made from three gray squares, each of area 25 cm², inside a larger white rectangle. Two of the vertices of the gray rectangle touch the midpoints of the shorter sides of the white rectangle and the other two vertices of the gray rectangle touch the other two sides of the white rectangle. What is the area, in cm², of the white rectangle?

(A) 125
(B) 136
(C) 149
(D) 150
(E) 172

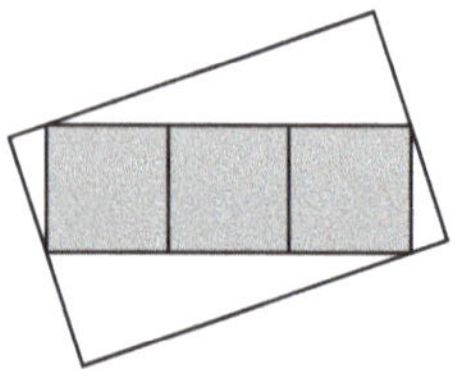

---

19 Angelo drew two lines meeting at a right angle. What is the smallest number of extra lines he could draw inside his right angle, as shown, so that for any of the values 10°, 20°, 30°, 40°, 50°, 60°, 70°, and 80°, a pair of lines can be chosen with the angle between them equal to that value?

(A) 2
(B) 3
(C) 4
(D) 5
(E) 6

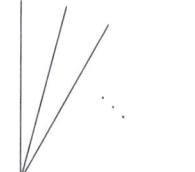

---

20 The sum of 2023 consecutive integers is 2023. What is the sum of the digits of the largest of these integers?

(A) 4
(B) 5
(C) 6
(D) 7
(E) 8

5 Points Each

21 Some beavers and some kangaroos are standing in a circle. There are three beavers in total, and no two beavers are standing next to each other. There are exactly three kangaroos who are standing next to another kangaroo. What is the largest possible number of kangaroos in the circle?

(A) 4
(B) 5
(C) 6
(D) 7
(E) 8

22. An ant is walking along the sides of an equilateral triangle. The speeds at which it travels along the three sides are 5 cm/min, 15 cm/min, and 20 cm/min, as shown. What is the average speed, in cm/min, at which the ant walks the whole perimeter of the triangle?

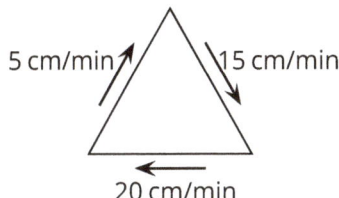

(A) 10
(B) $\frac{80}{11}$
(C) $\frac{180}{19}$
(D) 15
(E) $\frac{40}{3}$

23. Snow White organized a chess competition for the seven dwarves, in which each dwarf played one game with each of the other dwarves. On Monday, Grumpy played 1 game, Sneezy played 2, Sleepy 3, Bashful 4, Happy 5, and Doc played 6 games. How many games did Dopey play on Monday?

(A) 1
(B) 2
(C) 3
(D) 4
(E) 5

24. Sophia wants to write the numbers 1 to 9 in the regions of the shape shown so that the product of the numbers in any two adjacent regions is not more than 15. Two regions are said to be adjacent if they have a common edge. In how many ways can she do this?

(A) 12
(B) 8
(C) 32
(D) 24
(E) 16

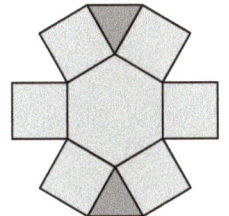

25. Martin is standing in line. The number of people in the line is a multiple of 3. He notices that he has as many people in front of him as behind him. He sees two friends, both standing behind him in the line, one in 19th place and the other in 28th place. In which position in the line is Martin?

(A) 14
(B) 15
(C) 16
(D) 17
(E) 18

26. Some mice live in three neighboring houses. Last night, every mouse left its house and moved to one or the other of the other two houses, always taking the shortest route. The numbers in the diagram show the number of mice per house, yesterday and today. How many mice used the path shown by the arrow?

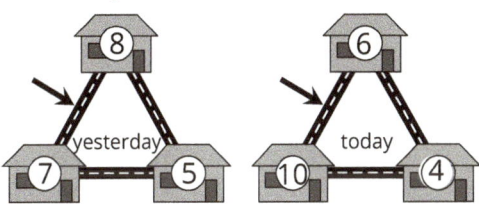

(A) 9
(B) 11
(C) 12
(D) 16
(E) 19

27. Bart wrote the number 1015 as a sum of numbers using only the digit 7. He used 7 a total of 10 times, as shown. Now he wants to write the number 2023 as a sum of numbers using only the digit 7, using a 7 a total of 19 times. How many times will he use the number 77?

(A) 2
(B) 3
(C) 4
(D) 5
(E) 6

```
  777
   77
+  77
   77
    7
─────
 1015
```

28 A regular hexagon is divided into four quadrilaterals and one smaller regular hexagon. The area of the shaded region and the area of the small hexagon are in the ratio $\frac{4}{3}$.
What is the ratio $\frac{\text{area of small hexagon}}{\text{area of big hexagon}}$?

(A) $\frac{3}{11}$
(B) $\frac{1}{3}$
(C) $\frac{2}{3}$
(D) $\frac{3}{4}$
(E) $\frac{3}{5}$

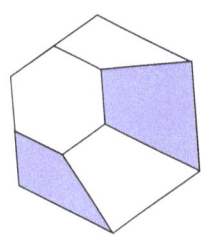

---

29 Jake wrote six consecutive numbers onto six white pieces of paper, one number on each piece. He stuck these bits of paper onto the top and bottom of three coins. Then he tossed these three coins three times. On the first toss, he saw the numbers 6, 7, and 8, as shown, and then colored them red. On the second toss, the sum of the numbers he saw was 23 and on the third toss the sum was 17. What was the sum of the numbers on the remaining three white pieces of paper?

(A) 18
(B) 19
(C) 23
(D) 24
(E) 30

30 A rugby team scored 24 points, 17 points, and 25 points in the seventh, eighth, and ninth games of the 2022 season. Their average of points per game was higher after 9 games than it was after their first 6 games. Their average after 10 games was more than 22. What is the smallest number of points that they could have scored in their 10th game?

(A) 22
(B) 23
(C) 24
(D) 25
(E) 26

2025

2025

3 Points Each

1 Lisa has four wooden digits. She can use them to form the number 2025. Which of the following numbers is the largest she can form with these digits?

(A) 2502
(B) 5202
(C) 5220
(D) 5502
(E) 5520

---

2 Isabelle rotates the hexagonal sheet of paper, as shown.

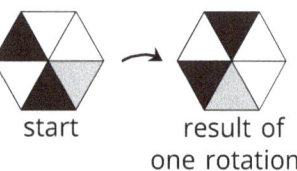

start   result of one rotation

Each rotation turns the hexagon through the same angle in the same direction. The figure shows the result of one rotation. Which of these numbers of rotations would leave the sheet looking as it did at the start?

(A) 7
(B) 8
(C) 9
(D) 10
(E) 12

---

3 Sandra rolls three dice and gets a total of 8. Each of the three dice shows a different number of dots. Which number of dots could Sandra not have rolled on any of her dice?

(A)
(B)
(C)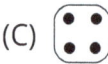
(D)
(E)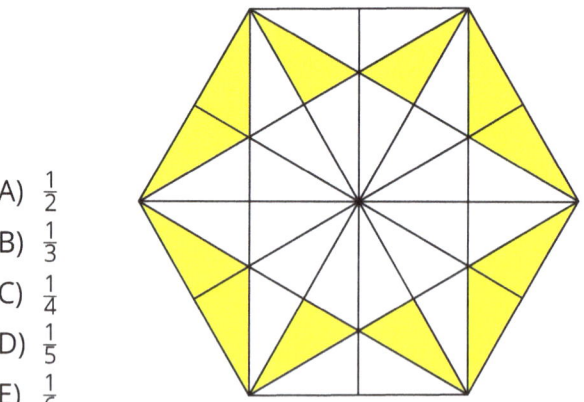

---

4 The regular hexagon shown is divided into many triangles of equal area. What fraction of the hexagon is shaded?

(A) $\frac{1}{2}$
(B) $\frac{1}{3}$
(C) $\frac{1}{4}$
(D) $\frac{1}{5}$
(E) $\frac{1}{6}$

---

5 How many sets of 12 minutes are there in 12 hours?

(A) 60
(B) 24
(C) 12
(D) 10
(E) 6

6 Daniel is 5 years old. His brother Dominic is 6 years older. What will the sum of their ages be in 7 years?

(A) 26
(B) 27
(C) 28
(D) 29
(E) 30

---

7 Omar wants to write the four digits 2, 0, 2, and 5 in the four boxes of the calculation shown. What is the smallest result that Omar could get?

☐ − ☐ + ☐ − ☐

(A) −7
(B) −6
(C) −5
(D) −4
(E) −3

---

8 There are ten more truth-tellers than liars in a room. Truth-tellers always tell the truth and liars always lie. Everyone in the room was asked, "Are you a truth-teller?" and everyone gave an answer. A total of 20 people answered, "Yes." How many liars are in the room?

(A) 0
(B) 5
(C) 15
(D) 20
(E) 25

---

9 Five circles, each with an area of 8 cm², overlap to form the figure shown. The area of each section where two circles overlap is 1 cm². What is the total area covered by the figure?

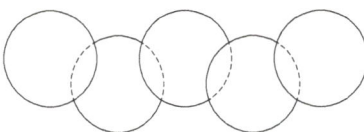

(A) 32 cm²
(B) 36 cm²
(C) 38 cm²
(D) 39 cm²
(E) 42 cm²

---

10 The real combination for the bicycle lock shown in the illustration is 0000. However, when someone looks at it from the side, they see 8888. When Paul looks at the combination of his friend's lock from the side, he sees 2815. What is the real combination of his friend's lock?

(A) 4037
(B) 4693
(C) 0639
(D) 0693
(E) 9603

4 Points Each

11 Matthew the mouse wants to get to a piece of cheese. He can only move horizontally or vertically between any two cells in the directions shown by the arrows. How many different routes can Matthew take to reach a piece of cheese?

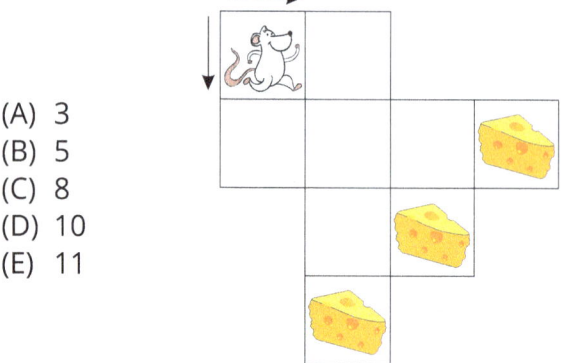

(A) 3
(B) 5
(C) 8
(D) 10
(E) 11

12 There are five hurdles in a 60 m hurdles race. The first hurdle is after 12 m. The gap between any two consecutive hurdles is 8 m. How far is the last hurdle from the finish?

(A) 16 m
(B) 14 m
(C) 12 m
(D) 10 m
(E) 8 m

13 Edgar wants to write a number in each circle in the diagram. He wants each number to be equal to the sum of the numbers in the two adjacent circles. He has already written two numbers, as shown. What number should he write in the gray circle?

(A) 2
(B) −1
(C) −2
(D) −3
(E) −5

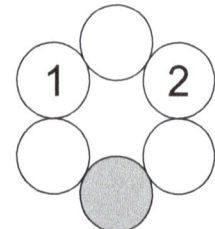

14 Louise places three rectangular pictures in the way shown. What is the value of x?

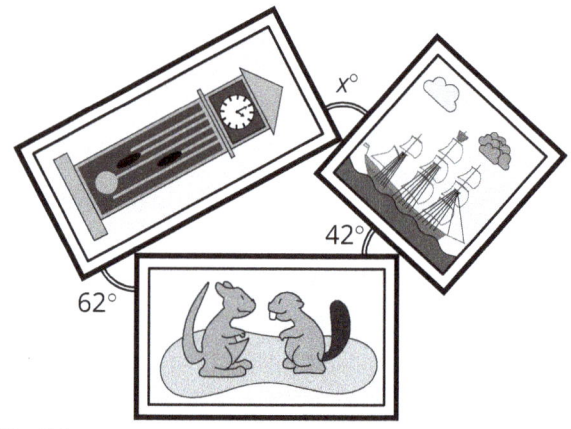

(A) 64
(B) 70
(C) 72
(D) 76
(E) 80

15 Warren is on a treadmill in the gym. He keeps looking at two stopwatches. The first shows the time elapsed since he started the session and the second the time remaining until the end of his session.

| 14:58 | 21:32 |

At some point the two stopwatches show the same reading. What do they show at that point?

(A) 17:50
(B) 18:00
(C) 18:12
(D) 18:15
(E) 18:20

16 Julia wants to fill in each □ with a different prime number less than 20 so that the value of A is an integer.

$$A = \frac{\square+\square+\square+\square+\square+\square+\square}{\square}$$

What is the maximum value of A?

(A) 20
(B) 14
(C) 10
(D) 8
(E) 6

17 Mark wants to fill in the cells on the diagram shown so that each cell contains either a cross or a circle. He also wants to ensure there is no line of four consecutive identical symbols in any column, row, or diagonal. When he has completed the diagram, what will the column colored gray contain?

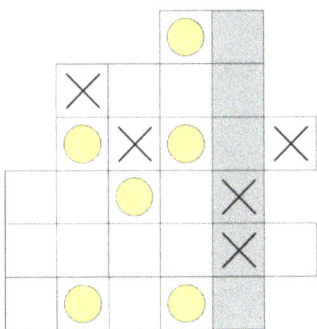

(A) 3 circles and 3 crosses
(B) 2 circles and 4 crosses
(C) 4 circles and 2 crosses
(D) 6 crosses
(E) a circle and 5 crosses

18 In the rectangle ABCD, the points E and F are marked on side DC as shown, so that ∠EBA = ∠DFA = 45° and AB + EF = 20 cm. What is the length of BC?

(A) 4 cm
(B) 6 cm
(C) 8 cm
(D) 10 cm
(E) 12 cm

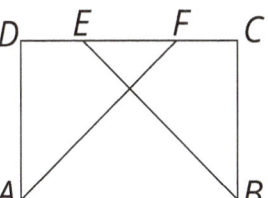

19 Sonia has two bowls of numbered balls. Bowl X contains seven balls numbered 1, 2, 6, 7, 10, 11, and 12. Bowl Y contains five balls numbered 3, 4, 5, 8, and 9. Which ball should Sonia transfer from Bowl X to Bowl Y to increase the average number on the balls in each bowl?

(A) 6
(B) 7
(C) 10
(D) 11
(E) 12

20 Peter drew a quarter circle with center at each corner of a flag with dimensions 12 cm by 9 cm and colored the region formed, as shown. What is the length indicated by the question mark?

(A) 5 cm
(B) 6 cm
(C) 7 cm
(D) 8 cm
(E) 9 cm

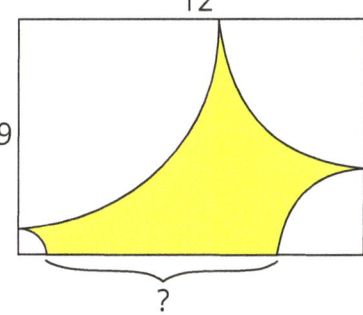

5 Points Each

21 In the six-digit integer *PAPAYA*, different letters stand for different digits and the same letter always represents the same digit. Also $Y = P + P = A + A + A$. What is the value of $P \times A \times P \times A \times Y \times A$?

(A) 432
(B) 342
(C) 324
(D) 243
(E) 234

---

22 During two sessions of soccer training, Paul shoots a total of 17 times at a target. He hits 60% of the shots he shoots in the first session. He hits 75% of the shots he shoots in the second session. How many times did he hit the target in the second session?

(A) 6
(B) 7
(C) 8
(D) 9
(E) 10

---

23 Anurag leaves for school at 8:00 a.m. His school is 1 km away. When he walks, his speed is 4 km/h. When he bikes, his speed is 15 km/h. He is 5 minutes early when he walks. How many minutes early is he when he bikes?

(A) 12
(B) 13
(C) 14
(D) 15
(E) 16

---

24 Ria draws four squares side by side, as shown. What is the area of the shaded quadrilateral?

(A) 54
(B) 60
(C) 66
(D) 72
(E) 80

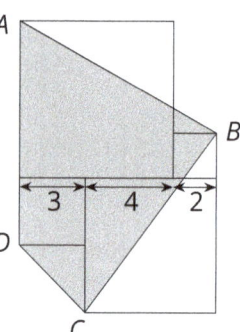

---

25 The letters *p*, *q*, *r*, *s*, and *t* represent five consecutive positive integers, though not necessarily in that order. The sum of *p* and *q* is 69 and the sum of *s* and *t* is 72. What is the value of *r*?

(A) 29
(B) 31
(C) 34
(D) 37
(E) 39

---

26 When the height of a cuboid is reduced by 3 cm, its surface area is reduced by 60 cm². The resulting shape is a cube. What is the volume of the original cuboid, in cm³?

(A) 75
(B) 125
(C) 150
(D) 200
(E) 225

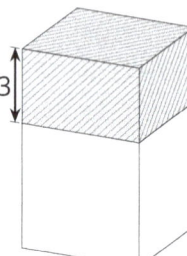

27 In the quadrilateral ABCD, the points N and K are marked on sides BC and AD respectively so that BN = 2NC and AK = KD. The area of triangle CKD is 2, and the area of triangle ABN is 6. What is the area of quadrilateral ABCD?

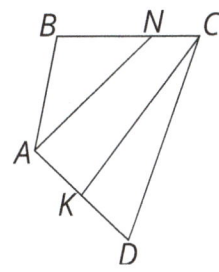

(A) 13
(B) 14
(C) 15
(D) 16
(E) 17

---

28 Some birds, including Ha, Long, Nha, and Trang, are perching on four parallel wires. There are 10 birds perched above Ha. There are 25 birds perched above Long. There are five birds perched below Nha. There are two birds perched below Trang. The number of birds perched above Trang is a multiple of the number of birds perched below her. How many birds in total are perched on the four wires?

(A) 27
(B) 30
(C) 32
(D) 37
(E) 40

---

29 The second figure below shows a net of an octahedron. Each face of the octahedron is divided into three parts. The octahedron is colored with the three colors, black, dark gray, and light gray, in such a way that the parts that come out of the same vertex or out of an opposite vertex are the same color. Which color could the part marked with the dot be colored?

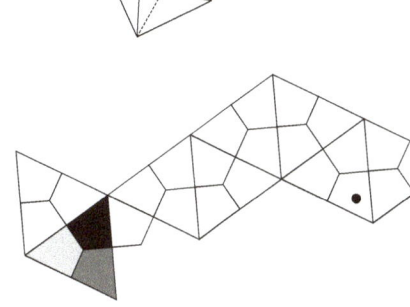

(A) Only black.
(B) Only dark gray.
(C) Only light gray.
(D) Both black and dark gray are possible.
(E) Both black and light gray are possible.

---

30 Adira keeps golden, red, black, pink, and white pearls in five small boxes. Each box contains pearls of only one color. The boxes are labeled as shown, and all the labels are true. Adira's friend Lilly wants to know which box contains the golden pearls. She may open exactly one of the five boxes to look inside. Which box must Lilly open to be certain which of the boxes contains the golden pearls?

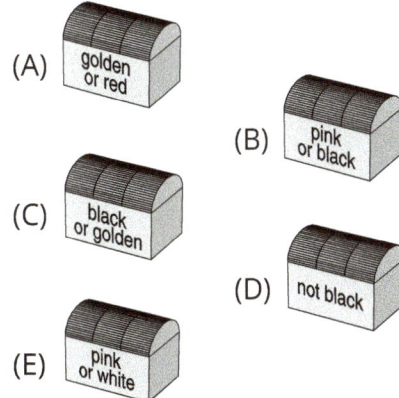

Part II
Solutions

2007

2007

3 Point Solutions

1 (C) 223
Perform long division with 2007 ÷ 9 = 223

2 (A) 22
Draw a line segment and count off ten equal segments marking dots. That gives 10 bushes on each side plus one bush at the end on each side which equals 22.

3 (E) 161
Rewrite the problem into minutes: 23 (1:37 to 2:00) + 60 (2:00 to 3:00) + 60 (3:00 to 4:00) + 18 (4:00 to 4:18) = 161.

4 (D) 27
We know that each die has a face with 1, 2, 3, 4, 5, and 6 dots on it. The sides which cannot be seen in the illustration of one cube are 3 + 4 + 5 + 6 and the other 1 + 3. Adding all the numbers together we obtain 27.

5 (D) 100 cm
Let r be the radius of the circle. There are 12 radii that make up the smaller rectangle. Solving $12r = 60$ gives $r = 5$. Therefore the diameter of each circle is 10 and the perimeter of the larger rectangle is 2(30) + 2(20) = 100.

6 (D) $\overline{CD}$
In order to be parallel to the x-axis, the line needs to be horizontal. Therefore, we need to find the two points where the y coordinates are the same, which are C and D.

7 (C) 34
Let c be the side of the smaller square. Using the Pythagorean Theorem we know that $c^2 = 5^2 + 3^2$. Therefore, c is equal to $\sqrt{34}$. Area = $(\sqrt{34})^2 = 34$.

8 (A) 2
Both of the numbers result in an odd number, because they are obtained by multiplying odd numbers by each other. Adding two odd numbers gives an even number. Therefore, the sum is divisible by 2.

9 (B) 989998
The biggest 6-digit palindromic number is 999999 and the smallest 5-digit palindromic number is 10001 The difference between them 989998.

10 (C) $\frac{1}{2}$
Set up the following equation and use the rules for exponents: $x = 2^{2006} \div 2^{2007} = 2^{-1} = \frac{1}{2}$.

4 Point Solutions

11 (C) $-2x$
Since x is a negative number, all of the expressions except $-2x$ will result in a negative number or possibly 0 for (A).

12 (C) 72 cm
Because all of the squares are placed on the segment AB, the sum of one side from all of the squares is equal to 24. Therefore, all of their perimeters will be equal to 24 × 3 = 72 cm since we cannot count the line itself.

13 (D) 16
Draw a diagram placing the points on the lines and draw segments connecting them. Count the triangles formed. From each of the two points on line b we get 6 triangles (3 small, 2 medium, and one large) and from each of the four points on line a we get 1 triangle. Altogether there are 16 triangles.

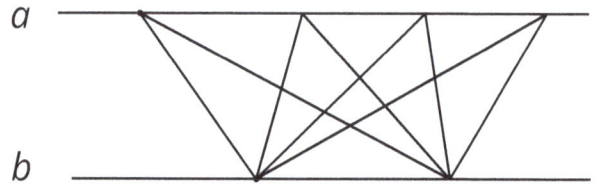

14 **(E) It will never stop.**
The Kangaroo will go in a continuous loop on the right side of the grid.

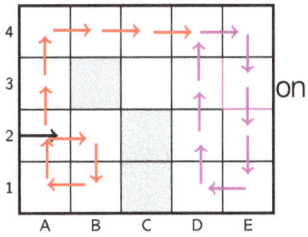

---

15 **(C) 6**
Set up the following equation by rewriting 9 as 3^2. Apply the rules for exponents and set them equal to each other to solve for x.

$$(3^3)^x = 9^9$$
$$(3^3)^x = (3^2)^9$$
$$3^{3x} = 3^{18}$$
$$3x = 18$$
$$x = 6$$

---

16 **(D) 40°**
$\angle BCD = 80° + 60° = 140°$
$\angle CBD = (180° - 140°) \div 2 = 20°$
$\angle ABD = 60° - 20° = 40°$

---

17 **(A) 1%**
The number of perfect squares in the set is 100 (since the square of 101 gives a value greater than 10 000). 100 out of 10 000 is 1%.

---

18 **(B) 2**
Since the unshaded region has two right angles which together add up to 180°, it covers half of the given square. The area of the square is 4, so the area of the shaded region is equal to 2.

---

19 **(B) 15**
Since we need one number from each row, let's break this down into three cases, using 1, 2, 3 from the top row. For each case we have two possibilities.
Case 1: We choose 1 + 5 + 9 = 15 or
    1 + 6 + 8 = 15.
Case 2: We choose 2 + 4 + 9 = 15 or
    2 + 6 + 7 = 15.
Case 3: We choose 3 + 4 + 8 = 15 or
    3 + 5 + 7 = 15.
In all the options, 15 is the sum.

---

20 **(E) 3**
Draw a diagonal line (the equation of the line is $y = x$). This line will intersect one of the small shaded boxes. Shade the three boxes on the right side forming a mirror image of the left side.

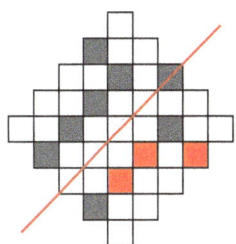

5 Point Solutions

21 **(D) 15**
There are five different patterns for inserting the two ones into 2007: There are 5 ways to write them next to each other. There are 4 ways to write them one number apart. There are 3 ways to write them two numbers apart. There are 2 ways to write them three numbers apart. There is one way to write them one on each side. Together we have 15 ways.

---

22 **(D) 8 km**

$$\text{speed} = \frac{\text{distance}}{\text{time}} \text{ so, time} = \frac{\text{distance}}{\text{speed}}.$$

a = distance traveled on flat ground
b = distance traveled up and also distance traveled down

$$\frac{a}{4} + \frac{b}{3} + \frac{b}{6} + \frac{a}{4} = 2$$

Simplifying we get $a + b = 4$. This mean that the length of one way trip is 4, which gives us 8 km for the entire trip.

---

23 **(E) 42**
To obtain the maximum number of cells we need a table with 7 horizontal lines and 8 vertical lines. The number of cells is a product of $(7 - 1)(8 - 1) = 42$.

24 (B) isosceles
Without any information about the lengths of the sides, we know that the only two triangles that are symmetrical along one of the angle bisectors are isosceles or equilateral. Since we know that the bisector from angle A divided the triangle into two congruent triangles, it was drawn from the vertex angle in an isosceles triangle. The triangle could be equilateral, but is not necessarily so.

---

25 (A) 2
Since n has two divisors, it is a prime number. Examine some numbers. Check that 2 cannot be an option. Let $n = 3$ and see that has 2 divisors. Now $n + 1 = 4$ has 3 divisors. Then $n + 2 = 5$, and that has 2 divisors again.

---

26 (C) 14
Use trial and error. Begin by assuming that Nick used some of the smallest numbers. $4 + 5 + 7 + 8 = 24$. If we can make a sum of $24 \times 3 = 72$ out of four of the five remaining numbers, we have found the answer. In fact, we can do this. $13 + 12 + 23 + 24 = 72$. The number that was not used is 14.

27 (C) 2

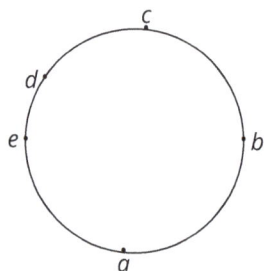

There are 5 points marked in the diagram. Because the question focuses on the sums of various combinations not being divisible by 3, we will focus on the reminders from division by 3. We consider this set: $\{0, 1, 2\}$, which represent the various integers by listing the remainders. Suppose we assume that none of the numbers have a remainder of 0. Then, because numbers 1 and 2 cannot be next to each other (because the sum would be divisible by 3), all numbers on the circle would be equal to 1 or all be equal to 2. However, this contradicts the conditions in our problem here, because the sum of three identical numbers (or three numbers with the same remainder when divided by 3) is divisible by 3. Therefore, at least one of the numbers is 0. Let's say $a = 0$. Then $b \neq 0$ and $e \neq 0$ and $b = e$. If $c \neq 0$ and $d \neq 0$ then $b = c = d = e$, which contradicts our problem, because again three numbers would give a sum that is divisible by 3. Then, either $c = 0$ or $d = 0$. Because c and d are neighboring numbers, they can't be both 0 at the same time. Now we can conclude that exactly two out of five numbers on the circle are divisible by 3.

---

28 (D) 110 dm

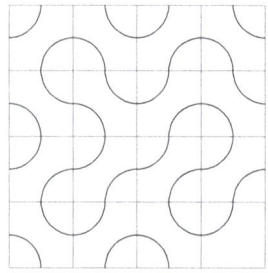

The illustration above is an example of such an unbroken curve of length 110 dm. It consists of 22 of the arc segments. Notice there is no other approach in which we can obtain a longer curve.

29. **(D) 5**

List the multiples of 9 from the set {100, ..., 999}. Call the set P = {108, 117, 126, ..., 999}. Then divide those numbers by 9 to obtain the set of numbers and again select the multiples of 9. Call that set Q = {45, 54, 63, 72, ..., 99, 108}. Notice that the sum of the digits in set Q is 9. Find corresponding numbers in set P where the digits add up to 18, so that the difference between the sum of the digits in the 3-digit number and in the number divided by 9 is 9. There are 5 such numbers: 486, 567, 648, 729, and 972. You can prove for all numbers what can be proved for the first one: 486. $4 + 8 + 6 = 18$, $486 \div 9 = 54$, $5 + 4 = 9$, and $18 = 2 \times 9$.

---

30. **(D) $2^6 \cdot 3^6 \cdot 5^4$**

Only $2^6 \cdot 3^6 \cdot 5^4$ can be obtained by starting with 15 and making 5 operations:
$2 \times 3 \times 5$, $2^2 3^2 5^2$, $2^3 3^2 5^2$, $2^3 3^3 5^2$, $2^6 3^6 5^4$.

2009

2009

3 Point Solutions

1 (A) 2009

$2009 > 200 \cdot 9 > 200 + 9 > 200 - 9 > 2+0+0+9$
$2009 > 1800 \;\;>\;\; 209 \;\;>\;\; 191 \;\;> 11$

---

2 (C) 2

All together the boys had $3 + 1 + 2 + 2 = 8$ dance partners. Since the first three girls had a total of 6 dance partners, the fourth girl must have had the remaining 2 dance partners.

---

3 (C) 18 cm

The white triangles form the perimeter of the star. There are 6 of them, and all are identical and equilateral, so the perimeter is the sum of the 12 edges and we know that it is 36 cm. Therefore, each edge is 3 cm. The shaded hexagon's perimeter is made of six of these 3 cm edges and is therefore 18 cm long.

---

4 (B) 20

The total number of pages from 15 to 53 is $53 - 15 + 1 = 39$. Since the page numbers he started and ended on are odd, there are 20 odd and 19 even numbered pages. So, he solved 20 problems by solving one problem on each of the odd numbered pages.

---

5 (D) $\frac{1}{324}$

The side length of the next smaller square is $\frac{1}{3}$ and the side length of the next smaller square is half of that, or $\frac{1}{6}$. Finally, the side length of the next smaller square, the black square, is a third of $\frac{1}{6}$, that is, $\frac{1}{18}$. To find the area we square the side length and since $18^2 = 324$, the area of the black square is $\frac{1}{324}$.

---

6 (D) 18

100 factors into prime factors as $2^2 5^2$. Using these prime factors and 1 the only four different numbers that multiply to 100 we can choose are 1, 2, 5, 10 and $1 + 2 + 5 + 10 = 18$.

---

7 (C) equal to half the number of dogs.

Let C be the number of cat paws and N the number of dog noses. We know that $C = 2N$ and we assume that each cat has 4 paws and each dog has one nose. So to obtain the number of cats, divide C by 4. Then the number of cats is $\frac{N}{2}$, which says that there are twice as many dogs as cats.

---

8 (C) 54°

Since $|PQ| = |PS|$ we know that $\triangle PQS$ is isosceles and therefore $2\angle PQS + 12° = 180°$ or $\angle PQS = \angle PSQ = 84°$. Then $\angle PSR = 96°$, and since $PS = SR$, the triangle $\triangle PSR$ is isosceles. Therefore, $2\angle SPR + 96° = 180°$ or $\angle SPR = 42°$. Finally, $\angle QPR = \angle QPS + \angle SPR = 54°$.

---

9 (C) 5

Let L be the number of large jars and S the number of small jars. Then $12L = 20S$ or equivalently $3L = 5S$. Having filled 9 large jars is therefore equivalent to having filled 15 small jars. Since Mother made enough juice to fill 20 smaller jars, she will need 5 more of the smaller jars.

---

10 (B) 2

We are looking for a pair (x, y) of integers such that $xy = x + y$. Clearly $(0, 0)$ is an example of such a pair and the only other possibility is $(2, 2)$.

4 Point Solutions

11 **(B) 3**
We only check the single digit positive integers, because for any double digit integer the square and cube differ at least by an order of magnitude. Only the integers 1, 2 and 4 have this property. $3^2 = 9$ but $3^3 = 27$ and for integers greater than 4 the square is a 2-digit number while the cube has 3 digits.

12 **(A) 3**
Remove the center dot and any other one and it is then obvious that at least one more must be removed, but removing any of the two diagonals solves the problem.

13 **(C) 45°**
The acute triangle cannot have an angle with a degree measure of 120, nor can any angle measure exceed 90 degrees. If we choose 55 and 10 to be the angle measures of the acute triangle, then the third angle would measure 115 degrees, which is impossible. The only other possibility are the angles with measure 80 and 55. Choosing these we see that the third angle measures 45 degrees and since all three angles are acute, the answer is 45 degrees.

14 **(A) $\frac{1}{4}$**
Draw the diagonals of the medium sized square so that each circle is now inscribed in a square. Then, looking just at one of the four smaller squares which contain a circle, notice that the reflection of the shaded region about the diameter of the circle shades exactly half of the square. Therefore, the shaded region is exactly a quarter of the total area.

15 **(D) 19**
If each cell contains a 1 then we have two possible sums: 10 and 19. Replacing one 1 with a 0 we now have the labels 19, 18, 10, 9. Introducing another 0 we have the labels 19, 18, 17, 10, 9, 8. Continuing thus we are led to believe that after inserting 9 zeros we have 20 labels, but this is incorrect as at this step we will have a first repeated label: 10 will appear twice. Thus, there are maximum of 19 possible different labels.

16 **(E) 24**
Two of the faces of the solid have the vertices (3, 6, a) and one has the vertices (a, a, 3). From here $2a + 3 = 3 + 6 + a$ and $a = 6$. The remaining 3 faces are (6, 6, b) and b must be 3. The sum of the vertices is $3 + 3 + 6 + 6 + 6 = 24$.

17 **(A) 1**
The answer has nothing to do with the given equation, but only depends on the fact that the correspondence between letters and digits is one-to-one and the crucial observation that there are 10 different letters. Thus, each letter is one of the digits between 0 and 9 and the product of any five digits is unique.

18 **(A) a**
$\left|\frac{1}{3} - \frac{1}{5}\right| = \frac{2}{15}$ and there are 16 smaller intervals so each is of length $\frac{1}{15 \cdot 8}$.

We need to solve the equation $\frac{1}{4} = \frac{1}{5} + \frac{k}{15 \cdot 8}$ where k is the number of notches we need to add to $\frac{1}{5}$.

$\frac{1}{4} = \frac{1}{5} + \frac{k}{15 \cdot 8} \iff 1 = \frac{4}{5} + \frac{k}{15 \cdot 2} \iff$
$\iff 1 = \frac{k}{3 \cdot 2} \iff k = 6$

The sixth notch corresponds to the letter a.

19 (E) 60°

Recall that the sum of the exterior angles is 360° and so each of the exterior angles measures 40°. The interior angles each measure $((9 − 2)180°) ÷ 9 = 140°$. Drawing the bisector of a we have two congruent triangles and enough information to compute the measure of half the angle a, which is 30°. So, $a = 60°$

---

20 (D) 92

Count the tiles that are needed to make a complete square: 5^2 for the first element, 6^2 for the second, 7^2 for the third and eventually $(4 + 10)^2 = 14^2$ for the tenth. Then subtract the 4 tiles that are missing and the shaded tiles. These are given by the pattern $1^2, 2^2, 3^2, \ldots$. Thus, to make the tenth element in the pattern we will need $14^2 − 4 − 10^2 = 196 − 4 − 100 = 92$ white tiles.

5 Point Solutions

21 (B) 13

If the first person in line is speaking the truth, then the last person in line is a liar. Then, what #25 said must be a lie, i.e., the person in front of the last person is a truth-teller. Consequently #23 is a liar, since #24, a truth-teller, spoke the truth. Continuing thus we see that #1 is a liar, a contradiction. Therefore, the first person in line is a liar. Therefore, the person behind the first person is a truth-teller, and spoke the truth. But #3 in saying that the person in front of #3 is a liar, told a lie, and is therefore a liar. Continuing thus we see that every odd number in line is a liar and so there are 13 of them.

---

22 (C) 64

We list all the possibilities:
$1 → 2 → 1$ or $3 → 2 → \ldots → 2$ accounts for $2^4 = 16$ different numbers and
$3 → 2 → 1$ or $3 → 2 → \ldots → 2$ accounts for $2^4 = 16$ different numbers and
$2 → 1$ or $3 → 2 → 1$ or $3 → \ldots → 1$ or 3 accounts for $2^5 = 32$ different numbers.
All together we can construct 64 different numbers with the desired property.

23 (C) $\frac{25}{76}$

Use the answer choices provided to solve this problem. Choices (A), (B), and (E) have a quotient more than $\frac{1}{3}$, and choice (A) also has a sum of 104. $\frac{25}{76} > \frac{25}{77}$, so the answer is $\frac{25}{76}$.

---

24 (D) 2:1

We observe that three cuts have been used and each cut was made parallel to two sides of the square. Thus, considering one cut at a time we observe that one cut introduces two more identical faces. The three cuts add six faces for a total of 12, or double the original amount. Thus the cuts double the original surface area.

---

25 (C) 2

If n is even then 2 is the smallest divisor, so the largest divisor must be 90 and $n = 180$. If n is odd and k is the smallest divisor of n, then k is odd and so is $45k$. Choosing $k = 3$, the greatest divisor is 135 and so $n = 405$. If $k > 3$ then the largest divisor is divisible by 5 and so is n, and therefore $k ≥ 7$ will not work. If $k = 5$, the largest divisor is 225, which is divisible by 3 and thus so is n, but $k = 5$ is supposed to be the smallest divisor. There are only the two possibilities we have listed.

---

26 (B) 45

If we want to minimize the side length of the large square we must minimize the side length of the 2009 smaller squares and all must be identical. Let each smaller square have side length 1. If there are 44^2 of them then the large square was divided into 1936 smaller squares, which is not enough, but $45^2 = 2025$ is more than enough. We can combine the surplus 16 squares into a single 4 × 4 square and thereby obtain $2025 − 16 = 2009$ squares.

27 (D) P, R, S

Consider first a quadrilateral PQRS where |PS| = |QR| = 2009 and |PQ| = |SR| = 2007. This is a rectangle so each angle measures 90 degrees. To obtain the quadrilateral in the question the point Q is shifted down and to the left. It is now at the intersection of two circles centered at P and R with radii 2006 and 2008, respectively. These circles intersect at two points, which in the rectangle PQRS are very close to either point Q or point S. In the second case we do not have a convex polygon and angle Q exceeds 180 degrees. Therefore, only angles P, R, S are necessarily less than 180 degrees.

---

28 (D) 40 cm²

The intersection of the areas of the two figures is $\frac{2}{3} \cdot 36 = 24$ cm². The area of the triangle is 60% of the area of the intersection so the area of the triangle must be 40 cm².

---

29 (D) Either C or D

We can easily color the first three rows and obtain

A	B	A	B	A
C	D	C	D	C
B	A	B	A	B
B				

Since both A and B border the shaded square only colors C and D can be used. The next row can be either CDCDC or DCDCD.

---

30 (C) 2

Let K be the point of intersection of the bisector of angle A and the side BC. Create an isosceles triangle by having L be the point on BC such that |BL| = |AB|. Then, since △ABC is isosceles we have ∠BAL = 80° = ∠BLA. From the information given it follows that also ∠AKL = 80°. Therefore, △AKL is isosceles, and |AK| = 2 = |AL|. We now know the measures of all remaining angles and we observe that △ALC is isosceles. Therefore, 2 = |AL| = |LC|. But |BC| − |AB| = |BC| − |BL| = |LC| = 2.

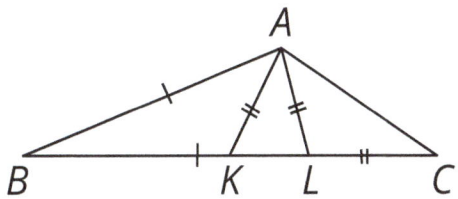

2011

2011

3 Point Solutions

1 **(D) 1 + 2011**

(A) $2011^1 = 2011$
(B) $1^{2011} = 1$
(C) $1 \times 2011 = 2011$
(D) $1 + 2011 = 2012$
(E) $1 \div 2011 = \frac{1}{2011}$

$2011 + 1 = 2012$ is the largest.

---

2 **(A) 42**
A cube has 6 sides and a tetrahedron 4. Together there are $5 \times 6 + 3 \times 4 = 42$ sides.

---

3 **(B) 7.5 m**
There are 8 white stripes and 7 black stripes. The crosswalk is $(7 + 8) \times (50\text{ cm}) = 7.5\text{ m}$ wide.

---

4 **(A) 2**
The calculator will compute the expression $(12 \div 3) - (4 \div 2) = 4 - 2 = 2$.

---

5 **(C) 50**
After 50 minutes the watch will display 21:01.

---

6 **(C) 12 cm²**
By drawing diagonals in the large square we see that the difference in the areas of the large and the medium squares is equal to twice the area of the small square.

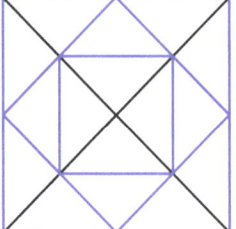

---

7 **(E) 21**
On the right side of the street the house numbers are 2, 4, 6, 8, 10, 12, so there are 6 houses on this side. On the left side of the street we have numbers 1, 3, 5, ... There are $17 - 6 = 11$ houses on the left side of the street, so these numbers are 1, 3, 5, 7, 9, 11, 13, 15, 17, 19, 21. So, last house on the left side is number 21.

---

8 **(C) 12**
1003, 1012, 1021, 1030, 1102, 1111, 1120, 1201, 1210, 1300, 2002, 2011, . . . , 4000.

---

9 **(C) 11**
We have $\frac{a+b}{2} = 17$ and $\frac{a+b+c}{3} = 15$. Then

$$\frac{45}{2} = \frac{a+b}{2} + \frac{c}{2}$$
$$\frac{45}{2} = 17 + \frac{c}{2}$$
$$\frac{11}{2} = \frac{c}{2}$$
$$11 = c.$$

---

10 **(B) 907**
The smallest such number is 107 and the largest is 800.

---

4 Point Solutions

11 **(C) 1**
The numerator is equal to $2011^2 \div 1000$ and the denominator is equal to $2011^2 \div (10 \times 100)$. Thus, the numerator and the denominator are equal.

---

12 **(E) 3–0**
The goal scored against the team had to be in the game they lost. This means that the tie was 0-0, and the team scored all three goals in the game they won. The results of the games that were lost, tied, and won were 0-1, 0-0, 3-0, respectively.

13 **(C) 3 g**
The 17 g necklace must have contained the 9 g and the 8 g pearls. The 13 g necklace had the 7 g and the 6 g pearls. The 7 g necklace could not have had the pearls weighing 4 g and 3 g, else it would not be possible to make the 5 g necklace using the remaining pearls. Thus, the 7 g necklace had the 5 g and the 2 g pearls, and the 5 g necklace had the 4 g and the 1 g pearls. The 3 g pearl remained unused.

---

14 **(B) 13**
The illustration shows one example of a path.

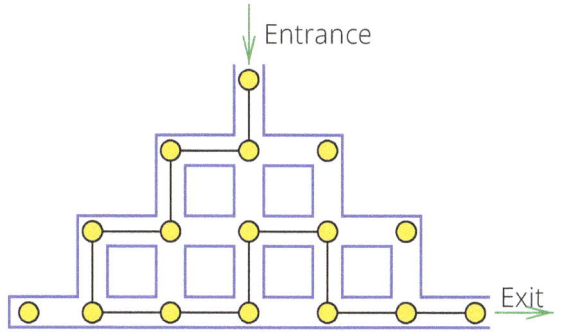

---

15 **(A) red**
The region directly above the one already colored red borders regions colored with three different colors. It must be colored orange. Similarly, the region below the one already colored green must be colored red. Now, the region to the right of the one already colored blue borders regions colored with three different colors. It must be colored green. The very top region must be colored blue, the bottom region orange and region X must therefore be colored red.

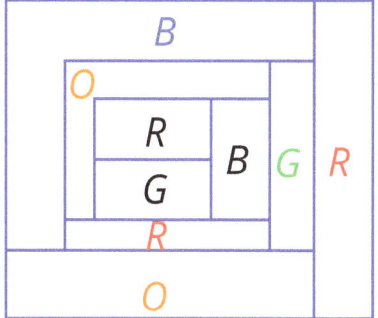

---

16 **(A) 5**
Peter caught 3, 4, and 5 fish on the three consecutive days, respectively.

---

17 **(E) 14 and 10**
The 8 numbers add to 96 so their arithmetic mean is 12. Find among them 6 numbers that add to 72, or equivalently remove two numbers that add to 24.

---

18 **(C) 6**
Four choices of the point F arise when DE is not the hypotenuse of the right triangle. If DE is the hypotenuse, then by a theorem of Thales, the point F will lie on a circle with diameter equal to 2. The equation of this circle, in Cartesian coordinates, is $x^2 + y^2 = 4$ and we also require that the area of our triangle be equal to 1, whence it follows that $xy = 2$. Substituting for y, we obtain the equation $x^4 - 4x^2 + 4 = 0$, which has exactly two distinct real-number solutions. These solutions correspond to the two distinct x-coordinates that give the final two possibilities for the point F.

---

19 **(B) $a + b$**
We have $0 < a < 1$ and $b > 1$. Therefore, $a \times b < b$ and $a \div b < 1$. Also, $b - a < b + a$ and $b + a > 1$, so $b + a$ is the largest of the listed quantities.

---

20 **(C) 3**

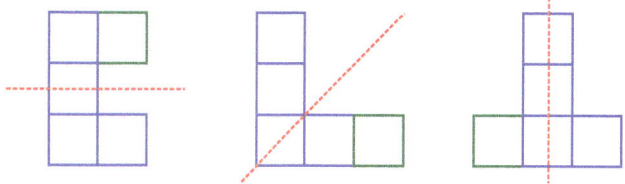

5 Point Solutions

21 **(D) 144 cm²**
Observe that, looking from left to right, there is one vertical line cutting through the whole square. This length is counted twice in the overall perimeter. Similarly, looking from top to bottom, there are two cuts that go across the entire square. These lengths are also counted twice. In addition to the edges of the square, there are 10 lengths, each equal to the side length of the square, that together make up the total perimeter of 120 cm. Thus, the square has an edge length equal to 12 cm and an area of 144 square centimeters.

22 **(E) 4**
Divisibility by 5 forces $y = 0$ or $y = 5$. Divisibility by 4 forces $y = 0$. Now, divisibility by 9 means the sum of the digits will be divisible by 9; i.e., the number $14 + x$ is divisible by 9. Thus, $x = 4$ and $x + y = 4$.

23 **(B)**
If shape (B) is placed as shown, the remaining regions have fewer than 5 squares each, so there is no room to place any of the other pieces.

24 **(D) 15 cm²**
Subtract from the area of the large square the areas of the two remaining squares. We have, in centimeters squared, $7^2 - 3^2 - 5^2 = 15$.

25 **(D) 20**
The 8-point shots have to come in multiples of three, or in groups of 24 points, so that together with a 5-point shot the score will be w 2-digit number ending in 9. Bob makes as many 10-point as 8-point shots, so assuming he hit the 8-point target 3 times, he will have earned 54 points with 6 shots. He needs some 5-point shots to get the score up to 99 points, while still missing 25% of the time. With 9 shots worth 5-points each he will have 99 points in 15 hits of the target, so since he missed a quarter of the time, Bob took 20 shots in all.

26 **(B) 15°**
Triangle CAB is isosceles, hence angle ACB measures 75° and angle ACD measures 65°. We now see that the triangle DAC is isosceles and, moreover, $|AD| = |AB|$. Since the triangle DAB is isosceles, angle ADB measures 50° and the measure of angle BDC is equal to 15°.

27 **(A)**
Observe that the bold line must intersect the vertex labeled C and also pass through the middle of the edge AD. This eliminates all but one possibility.

28 **(B) 2**
After some cancellation we have
$$\frac{K \times N \times A \times R \times O \times O}{M \times E}$$
Pick numbers to make the numerator as small as possible and the denominator as large as possible. Choosing $\frac{2 \times 3 \times 4 \times 6 \times 1 \times 1}{8 \times 9}$ yields 2.

29. (B) 37

From the illustration we see that the length of side x is equal to the width of the shape of Figure 1 plus 13, or 11 + 13 + 13 = 37.

---

30. (B) John

Let the points A, J, K represent the respective location of Adam, John, and Karl, and plot these points on a number line. Using absolute value, we can express each of their statements in the following way.

1. Adam: $|A - J| > 2|A - K|$
2. John: $|J - K| > 2|J - A|$
3. Karl: $|K - J| > 2|K - A|$

Since at least two statements are true, we check the consistency of the above equations two at a time. Suppose first that both Adam and John are telling the truth. Adding inequalities (1) and (2), we have

$$|A - J| + |J - K| > 2|A - K| + 2|J - A|$$
$$> 2|J - K|,$$

by the triangle inequality, whence $|A - J| > |J - K|$. But, this together with John's statement yields $|A - J| > 2|A - J|$, which is a contradiction. Next, suppose that both John and Karl are telling the truth. Adding inequalities (2) and (3), we have

$$|J - K| + |K - J| > 2|J - A| + 2|K - A|$$
$$> 2|J - I|,$$

by the triangle inequality, whence we arrive at the contradiction $|J - K| > |J - K|$. Finally, suppose that both Adam and Karl are telling the truth. Adding inequalities (1) and (3), we have

$$|A - J| + |K - J| > 2|A - K| + 2|K - A|$$
$$= 4|A - K|,$$

which does not lead to a contradiction. We conclude that John's statement is false.

2013

2013

3 Point Solutions

1 (D) 6
Divide all three shaded parts of a triangle into two equilateral triangles. In this way, 6 small triangles are shaded out of 9 equilateral triangles that are part of the big triangle. Calculating the area of the shaded part as $\frac{6}{9}$ of 9 gives 6.

---

2 (D) 55
It is given that $\frac{1111}{101} = 11$. Notice that

$$\frac{3333}{101} + \frac{6666}{303} = 3\left(\frac{1111}{101}\right) + \frac{6}{3}\left(\frac{1111}{101}\right) =$$
$$= 3\left(\frac{1111}{101}\right) + 2\left(\frac{1111}{101}\right) = 5\left(\frac{1111}{101}\right) = 5(11) = 55.$$

---

3 (A) 35
It is given that $\frac{\text{mass of salt}}{\text{mass of fresh water}} = \frac{7}{193}$.
There are 1000 kg of sea water. Let x be the mass of salt and $1000 - x$ be the mass of fresh water. Then the following equation may be expressed: $\frac{7}{193} = \frac{x}{1000-x}$. Solving for x gives the mass of salt in 1000 kg of sea water:

$$\frac{7}{193} = \frac{x}{1000-x},$$
$$7(1000 - x) = 193x,$$
$$7000 - 7x = 193x,$$
$$200x = 7000,$$
$$x = 35 \text{ kg}.$$

---

4 (C) 4
The illustration below shows that 4 is the smallest number of cells remaining after cutting out the desired shapes.

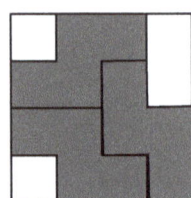

---

5 (E) 11
The smallest numbers that Roo can choose from are 2 digit numbers whose products are 24: 38, 83, 46, 64. The smallest out of these four numbers is 38. The sum of its digits is 11.

---

6 (E) 6
There are 5 different colors of the balls in the bag. In order to make sure that two balls of the same color are taken we must take out at least 6 balls. Out of the 6 balls at least 2 of them must be of the same color.

---

7 (C) 4
Within 55 minutes Alex lights all 6 candles. The first candle she lights at the beginning, the second after 10 minutes, the third after 20 minutes, the fourth after 30 minutes, the fifth after 40 minutes, and the sixth after 50 minutes. Since each candle burns for 40 minutes, the first and the second candles will burn out before 55 minutes is up. This means only 4 candles are still lit.

---

8 (E) 2.5
In order to find the average number of children in five families, we must divide the total number of children by 5. Notice that all numbers but one can be expressed as a quotient of a number and 5. (A) $0.2 = \frac{1}{5}$, (B) $1.2 = \frac{6}{5}$, (C) $2.2 = \frac{11}{5}$, (D) $2.4 = \frac{12}{5}$. Only answer (E) 2.5 cannot be the average number of children in five families since $2.5 = \frac{25}{10} = \frac{12.5}{5}$. It is impossible to have a total number of children equal to 12.5.

---

9 (A) 4
Mark's speed is 9/8 of Liza's speed. This means Mark's speed is 1/8 greater that Liza's. If they start from the same point, Mark would catch up with Liza after 8 circuits. However, they are only half a circle apart. This means it would take Mark only half of the distance, which is 4 circuits, to catch up with her.

10 (C) 14
In order to fulfill $x \times y = 14$, $y \times z = 10$, and $z \times x = 35$, x must be equal to 7, y must be equal to 2, and z must be equal to 5. Then the value of $x + y + z$ is calculated as $7 + 2 + 5 = 14$.

4 Point Solutions

11 (E) 8
The illustration below shows 8 possible positions of a 3 × 1 ship.

12 (E) 130°
Look at the diagram below. Notice that angles x and y can be easily found by the triangle angle sum theorem. Knowing $α = 55°$ and $β = 40°$, we can calculate $x = 85°$. Knowing $β = 40°$ and $y = 35°$, we calculate $y = 105°$. In order to find angle $δ$, we must notice that $DBCF$ is a quadrilateral, and the sum of the angles in any quadrilateral is always 360°. Knowing angles x, y, and $β$, we calculate $δ = 360° − (85° + 105° + 40°) = 130°$.

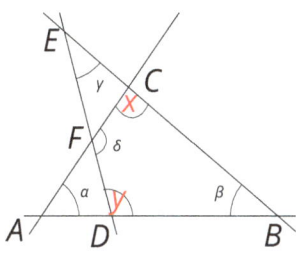

13 (B) 60° and 60°
The only trapezoid with sides as integers and a perimeter equal to 5 has the sides equal to 1, 1, 1, and 2 as shown in the diagram. Notice that the shaded part of the trapezoid is a special right triangle with angles 30-60-90 (because it's a right triangle with the hypotenuse equal to 1 and the shorter leg equal to half of the hypotenuse). The acute angle that is a part of the trapezoid is equal to 60° and so is the other angle since it is an isosceles trapezoid.

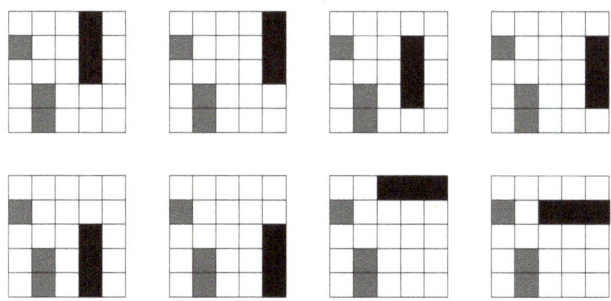

14 (C)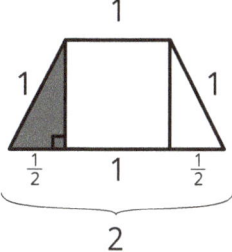
Only if net (C) is folded, a cube cannot be formed. The lower part of the net, the two triangles, would overlap and not form a square.

15 (B) 45
(A) Example of 40% of odd numbers among several consecutive numbers: 2, 3, 4, 5, 6. There are 2 odd numbers out of 5 consecutive numbers. $\frac{2}{5} = \frac{40}{100} = 40\%$

(B) 45% of odd numbers among several consecutive numbers cannot be achieved. Note that $45\% = \frac{45}{100} = \frac{9}{20}$. This means you would have to have 9 odd numbers and 11 even numbers out of 20 consecutive numbers. This case is not possible because the number of odd and even numbers in any set of several consecutive numbers can differ by at most 1. 9 and 11 differ by 2.

(C) Example of 48% of odd numbers among several consecutive numbers: 2, 3, 4, 5, 6, 7, 8, 9, 10, 11, 12, 13, 14, 15, 16, 17, 18, 19, 20, 21, 22, 23, 24, 25, and 26. There are 12 odd numbers out of 25 consecutive numbers. $\frac{12}{25} = \frac{48}{100} = 48\%$

(D) Example of 50% of odd numbers among several consecutive numbers: 1, 2, 3, 4, 5, 6, 7, 8, 9, and 10. There are 5 odd numbers out of 10 consecutive numbers. $\frac{5}{10} = \frac{50}{100} = 50\%$

(E) Example of 60% of odd numbers among several consecutive numbers: 3, 4, 5, 6, and 7. There are 3 odd numbers out of 5 consecutive numbers. $\frac{3}{5} = \frac{60}{100} = 60\%$

16 (A) A

Rectangle ABCD is located in the 4th quadrant. This gives x a positive value and y a negative value. A quotient of these two numbers is negative. We obtain the least value by dividing the least value of y by the least value of x, which is point A.

17 (A) 702

Numbers in the increasing order: 1023, 1032, 1203, 1230, 1302, 1320, 2013, 2031, 2103, 2130, 2301, 2310, 3012, 3021, 3102, 3120, 3201, 3210. Notice that out of all these numbers, we can assume neighboring numbers will have the largest difference in case of 1320 and 2013 or 2310 and 3012. The difference between 1320 and 2013 is 693. The difference between 2310 and 3012 is 702. Out of these two differences 702 is the greatest.

18 (E) 32

Look at the diagram below. There are 32 cells that are not intersected by either diagonal of a 6 × 10 grid.

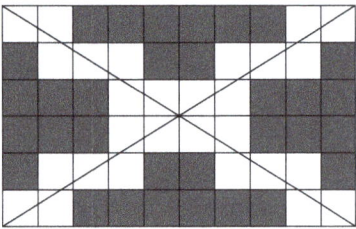

19 (B) Betty

Betty is the only one who was not born on the same day as another child, so she was born on 04/23/2001. This is also the latest birth date listed, so she is the youngest.

20 (C)

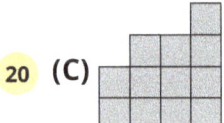

When John looks from the back, he sees from left to right the following number of blocks: 2, 3, 3, and 4, as the highest number of blocks visible. The correct answer is (C).

5 Point Solutions

21 (B) 84 cm²

Look at the diagram below. The area of the quadrilateral KLMN can be calculated by finding the area of rectangle ABCN and subtracting four areas from it: triangle LBM, triangle MCN, triangle DKN, and trapezoid ALKD. Considering that each cell of the grid has sides of length 2 cm, we calculate the area of rectangle ABCN = 14 × 10 = 140 cm². Area of LBM = $\frac{1}{2}$ × 6 × 8 = 24 cm². Area of MCN = $\frac{1}{2}$ × 2 × 14 = 14 cm². Area of DKN = $\frac{1}{2}$ × 2 × 8 = 8 cm². Area of ALKD = $\frac{1}{2}$(8 + 2) × 2 = 10 cm². The area of KLMN = 140 − (24 + 14 + 8 + 10) = 140 − 56 = 84 cm².

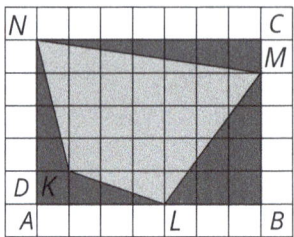

22 (D) S = 2013Q

Note that according to the properties of exponents, $2013^6 = (2013^2)^3$, and also $2013^6 = (2013^3)^2$. Therefore, among the integers from 1 to 2013^6, there are 2013^2 cubes and 2013^3 squares. Since $2013^3 = 2013 \times 2013^2$, then S (the number of squares) equals 2013Q (the number of cubes).

23 (C) 23
Notice that the last digit of the sum of the 5-digit and the 4-digit numbers is 3. This means that the last digits in both numbers are different, which implies that the last digit of the first number was deleted. In order to get 3 as the last digit of the sum of the two numbers you must add 2 with 1, 1 with 2, 9 with 4 or 4 with 9. Working backwards with the addition of the 5-digit number and the 4-digit number, we find that only if the last digit in the first number is 1 and the last digit in the second number is 2 we get the sum equal to 52,713. The original 5-digit number is 47,921 and the 4-digit number is 4,792. The sum of the original 5-digit number is 23 (4 + 7 + 9 + 2 + 1).

---

24 (C) 12
The greatest number of maples that the gardener can plant with the number of trees between any two maples not equal to three is 12. Plant 4 maple trees, then 4 linden trees, 4 maple trees, 4 linden trees, and 4 more maple trees.

---

25 (B) 41
The following scenario takes place in the problem: Andrew finished ahead of twice as many runners as finished ahead of Daniel. Andrew finished ahead of 20 runners, and there were 10 runners ahead of Daniel. 20 is twice as much as 10. Daniel finished ahead of 1.5 times as many runners as finished ahead of Andrew. Daniel finished ahead of 30 runners, and there were 20 runners ahead of Andrew. 30 is 1.5 times greater than 20. Finally, Andrew finished in 21st place as shown in the diagram.

10 runners ← Daniel ← 9 runners
　　　　　　← Andrew ← 20 runners

---

26 (A) 9
Look at the illustration below. Let's mark all four cars with numbers: 1, 2, 3, and 4 and distinguish all 4 directions as A, B, C, and D. Nine following scenarios as to where each car leaves can take place:
1. 1B-2C-3D-4A
2. 1B-2D-3A-4C
3. 1B-2A-3D-4C
4. 1D-2A-3B-4C
5. 1D-2A-3C-4B
6. 1D-2C-3D-4A
7. 1C-2A-3D-4B
8. 1C-2D-3A-4B
9. 1C-2D-3B-4A

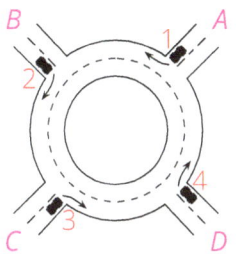

---

27 (B) −671
By writing several terms of the sequence, a pattern can be observed: 1, −1, −1, 1, −1, −1, 1, −1, −1, 1, −1, −1, 1, −1, −1, 1, −1, −1, ... Notice that consecutive three terms of the sequence add up to −1. There are 671 sets of three numbers in 2013. Check this by dividing 2013 by 3. Multiplying 671 by −1 we get a number equal to −671.

---

28 (D) 456231
In this situation, at first, 4 pies are in the kitchen. Kids eat pie number 4 as the hottest. After the 5th pie is baked they eat it as the hottest. After the 6th pie is baked they eat it as the hottest. Now in the kitchen there are 3 pies left: 1st, 2nd and 3rd. The hottest is the third, as baked last out of the three remaining. So they would have to go for the 3rd pie before the 2nd as shown in the order.

29 (B) 5

Number 5 on edge RS will satisfy this rule. Proof: Let p, q, r, s be the numbers corresponding to the four vertices. There are 3 pairs (mutually disjoint) of non-adjacent edges: PQ and RS, QR and PS, RP and QS. The sum of the numbers assigned to each of the 3 pairs is $p + q + r + s$. Thus, the sum of all (different) numbers assigned to 4 vertices and 6 edges is $(p + q + r + s) + 3(p + q + r + s) = 4(p + q + r + s)$ and it is equal to $1 + 2 + 3 + 4 + 5 + 6 + 7 + 8 + 9 + 11 = 56$, so $p + q + r + s = 56 \div 4 = 14$. Since it's given that $p + q = 9$, then $r + s = 5$, which is the number assigned to the edge RS. The assignement can actually happen as shown in the diagarm below.

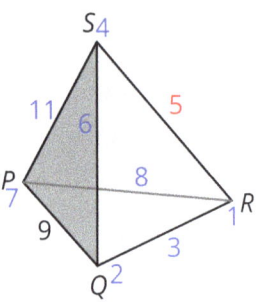

30 (C) All such integers N are divisible by 6.

Find counterexamples to eliminate answers. Notice that statement (A) is not true because 30 is an integer not divisible by 4 that is smaller than the sum of its three greatest divisors: $30 < (15 + 10 + 6)$. (B) is not true because 24 is an integer not divisible by 5 that is smaller than the sum of its three greatest divisors: $24 < (12 + 8 + 6)$. (D) is not true because 36 is an integer not divisible by 7 that is smaller than the sum of its three greatest divisors: $36 < (18 + 12 + 9)$. (E) is not true because of the examples in (A), (B), or (D). All such integers are divisible by 6.

2015

3 Point Solutions

1 (A) $\frac{2+0+1+5}{1+5}$

Checking expression (A), we find
$\frac{2+1+5}{1+5} = \frac{8}{6} = \frac{4}{3}$, which is desired.

---

2 (A) 1 hour and 35 minutes
2 hours and 10 minutes = 130 minutes,
130 minutes – 35 minutes = 95 minutes,
and 95 minutes = 1 hour and 35 minutes.

---

3 (B) 20 cm
Let ℓ be the length of a small rectangle, and w be its width. Then $\ell = 10$, and $w = 5$. Therefore the length of the longer rectangle is $\ell + 2w = 20$ cm.

---

4 (E) 1000
2.015 × 510.2 is slightly more than 2 × 510 = 1020, so 1000 is closest to the product.

---

5 (A) 4, 6, 11
If you imagine the cube being folded together, you will see that 1 is opposite 3, 2 is opposite 4, and 5 is opposite 6. Therefore, the three sums will be 1 + 3 = 4, 2 + 4 = 6, and 5 + 6 = 11.

---

6 (D) $\frac{2014}{4}$

Clearly, (A) is an integer. 2012 is divisible by 2, so (B) is an integer as well. The same is true for (C) and (E). However, (D) is not an integer because our divisibility rules for 4 tell us to check if 14 is divisible by 4.

---

7 (D) 9
The perimeter of the original triangle is 6 + 10 + 11 = 27. An equilateral triangle has all equal sides, so its side length is 27 ÷ 3 = 9.

---

8 (C) *XY*
First, fold up the triangle *WUV* (along *WU*) and the triangle *SQR* (along *SQ*). Next, fold up the rectangle *WSTU* (along *WS*) and push it toward the triangle *SRQ*, so *ST* and *SR* form one edge of the prism. At the same time *WV* and *WX* become one edge. Then fold up the rectangle *XQPY* and push toward the edge *TU*, so *XY* and *UV* become one edge.

---

9 (C)

The largest area Simon can cover is inside the circle of radius 5 with the center at the trunk and outside the circle of radius 5 with the center at the doghouse. (B) is the overlap of these circles and that overlap must be removed from Simon's circle, so (C) is the answer.

---

10 (D) 20
The cyclist travels 5 m, or 500 cm per second. This means that each wheel rotates 500 cm per second. Since each wheel has a circumference of 125 cm, we can say that each wheel rotates 500 ÷ 125 = 4 times per second. Thus, in five seconds each wheel will rotate 4 × 5 = 20 times.

4 Point Solutions

11 (B) 19
We know that if a new boy or girl enters the class, the condition for that gender of students will be violated. Therefore, there is one boy born every day of the week and one girl born in every month. This gives us a total of 7 + 12 = 19 children in the class.

---

12 (C) 1
Note that the white triangle in the topmost square is equal to the gray triangle in the bottom right triangle. Thus the shaded area is 1.

---

13 (B) 2
In our equation, we can assume that every asterisk before a 0 is a minus, since it makes no difference whether you add or subtract a 0. So we are left with 2 * 1 * 5 * 2 * 1 * 5 * 2 * 1 * 5 = 0. Our goal is to split up the remaining numbers into two groups with equal sums, with one having as few numbers as possible. We can then assign pluses to the group with fewer numbers and minuses to the group with more numbers. The sum of all the numbers is 24, and we can form the two groups as (2, 5, 5) and (1, 2, 1, 2, 1, 5). The first 2 does not need a plus sign, so we can just use two plus signs. One example of a solution is 2 − 0 − 1 − 5 − 2 − 0 − 1 + 5 − 2 − 0 − 1 + 5 = 0.

---

14 (D) 1.5 cm
Since one liter equals 1000 cubic cm, we have 15000 cubic cm per square meter. To find how much the water level rises in the pool, we must divide by 1 square meter, which is equivalent to dividing by 10000 square cm. 15000 ÷ 10000 = 1.5

---

15 (E) None of (A) to (D).
Let x be the number of branches with leaves only. Then L, the total number of leaves, is
$5x + 2(10 − x) = 3x + 20$
So, $L − 20$ must be divisible by 3. None of the choices are divisible by 3 when 20 is subtracted, so none of them work.

Or:
Make a table.
The first colum lists the number of branches with 5 leaves, the second branches with 2 leaves, and the last the total. None of the listed answer choices work.

10	0	10 × 5 + 0 × 2 = 50
9	1	9 × 5 + 1 × 2 = 47
8	2	8 × 5 + 2 × 2 = 44
7	3	7 × 5 + 3 × 2 = 41
6	4	6 × 5 + 4 × 2 = 38
5	5	5 × 5 + 5 × 2 = 35
4	6	4 × 5 + 6 × 2 = 32
3	7	3 × 5 + 7 × 2 = 29
2	8	2 × 5 + 8 × 2 = 26
1	9	1 × 5 + 9 × 2 = 23
0	10	0 × 5 + 10 × 2 = 20

---

16 (C) 3
Let x equal the mean score of those who failed the test: $0.6 \cdot 8 + 0.4x = 6$ and thus $x = 3$.

17 (C) 8

The area of the corner triangle is the difference between the area of the square and the area of the pentagon. Therefore, the triangle has the area 1 since two consecutive integers differ by 1. It is geometrically obvious that the area of the corner triangle is $\frac{1}{2}$ of a quarter of the original square. So the area of the triangle is $\frac{1}{8}A$ where A is the area of the square. Thus, $\frac{A}{8} = 1$ and $A = 8$.

---

18 (B) 56 cm

Let x and y be the side lengths of the rectangle with $x > y$. Then $2x + y = 44$ and $x + 2y = 40$. Adding equations gives $3(x + y) = 84$. Thus the perimeter is $2(x + y) = 2 \times (84 \div 3) = 56$.

---

19 (A) only green

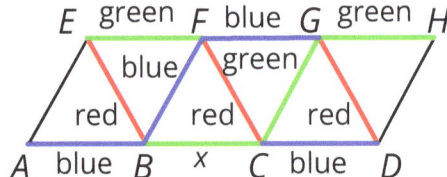

The vertices are labeled A through H. In $\triangle ABE$ the side BE can't be blue and in $\triangle BFE$ the same side can't be green, so BE is red. Therefore, the side BF is blue. Similarly, the side DG (look at $\triangle CDG$ and $\triangle DHE$) is red, so the side CG is green. For the same reason the common side, CF, of $\triangle BCF$ and $\triangle CGF$ must be red. Consequently, BC is green.

---

20 (B) 1

Let n be the number of children that studied. Note that if $n = 0$, then Pol is right, which means that he didn't study, so he is not telling the truth. Thus $n > 0$. It follows that exactly n children will tell the truth, saying that n children studied. Exactly 1 child, Berta, said that 1 child studied. So we know that only Berta studied.

---

5 Point Solutions

21 (C) 6

Notice that all neighbors of the region with −4 are neighbors of the region with 2. The only additional neighbor of the region with 2 is the central region with "?". Therefore, 2 = −4 + ?. It follows that ? = 2 + 4 = 6. One possibility is shown on the diagram below.

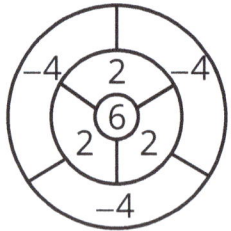

---

22 (A) 6

We know that 25% of the weight is taken up by the two lightest kangaroos and 60% of the weight is taken up by the heaviest three kangaroos. The rest of the weight is 15%, which is less than the weight of the two lightest kangaroos. Therefore, there can only be one extra kangaroo because if there were two or more, they would weigh less than the two lightest. So the total number of kangaroos is 2 + 1 + 3 = 6. One possible weight distribution is 12%, 13%, 15%, 19%, 20%, and 21% of the total weight.

---

23 (B) 30°

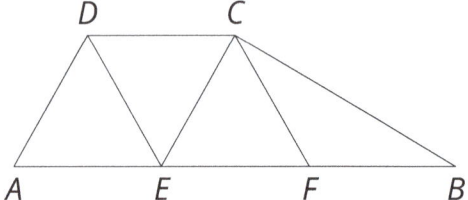

Consider the above image. In the trapezoid CD = DA = AE = EF = FB, so AB = 3CD. $\triangle DAE$, $\triangle ECD$, and $\triangle EFC$ are congruent equilateral triangles, so $\angle CFE = 60°$, and thus $\angle CFB = 120°$, and FC = FB, so $\angle ABC = \angle FBC = 30°$.

24 (C) 48

Let $c_1 \le c_2 \le c_3 \le c_4 \le c_5$ be integers written on the 5 cards. Then $c_1 + c_2 = 57$ and $c_4 + c_5 = 83$. 57 and 83 are odd, so $c_1 \ne c_2$ and $c_4 \ne c_5$. Thus $c_1 < c_2 \le c_3 \le c_4 < c_5$. If $c_2 < c_3$, then $c_1 < c_2 < c_3 \le c_4 < c_5$ and $c_1 + c_2 < c_1 + c_3 < c_2 + c_3$, so we would have more than three sums. Therefore $c_2 = c_3$.
If $c_3 < c_4$, then $c_1 < c_2 \le c_3 < c_4 < c_5$ and $c_1 + c_2 < c_1 + c_4 < c_2 + c_4 < c_2 + c_5$, so we would have more than three sums. Therefore $c_3 = c_4$ and $c_1 < c_2 = c_3 = c_4 < c_5$.
Consequently, $2c_2 = c_2 + c_3 = 70$, $c_4 = c_3 = c_2 = 35$, $c_1 = 57 - 35 = 22$, and $c_5 = 83 - 35 = 48$.

25 (C) 671

We see that as long as the quotient is 2, decreasing the divisor by 1 increases the remainder by 2. $2015 = 2 \times 1000 + 15$, $2015 = 2 \times 999 + 17$, $2015 = 2 \times 998 + 19 \ldots$. Since $2015 = 671 \times 3 + 2$, we know that the last time the integer quotient is 2 happens for 672. $2015 = 2 \times 672 + 671$, so 671 is the highest remainder since $2015 = q \times d + r$, $0 \le r < d$ and $m \le 671$ imply $r < 671$. Therefore, the greatest remainder we can have for this problem is 671.

26 (D) d

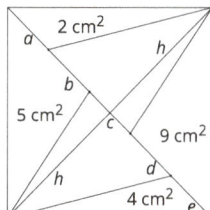

The two vertices of the square that are not the endpoints of the diagonal are equally distanced from the diagonal. Let h be the distance, so h is the height in all six triangles. The triangle of the segment $b + c$ has the area $30/2 - 2 - 9 = 4$ and the triangle of the segment $c + d$ has the area $30/2 - 5 - 4 = 6$. For each of the six triangles, the area is $\frac{1}{2} \times h \times$ base, so base $= 2 \times$ area$/h$. See the table.

Area	2	5	4	6	9	4
Base	a	$a+b$	$b+c$	$c+d$	$d+e$	e
Length	$4/h$	$10/h$	$8/h$	$12/h$	$18/h$	$8/h$

Thus $a = 4h$, $b = 6h$, $c = 2h$, $d = 10h$, $e = 8h$. So d is longest.

27 (D) 6

Suppose 1 is green. Either there is no other green number or there is a green number greater than 1.
(Case 1): If there is no other green, then the pattern is GRRRRRRRR...
(Case 2): If there is another green number, then look at the first such number. Call it g. Then, since g and 1 are green, we know that only numbers greater than 1 and less than g are red. Consider the cases when $g = 2$, $g = 3$, and $g \ge 4$.
(i) $g = 2$, so the pattern is (green only) GGGGGGGGGGGGGGG...
(ii) $g = 3$, so only 2 is red. The pattern is GRGGGGGGGGGGGGG...
(iii) $g \ge 4$, so 2 and 3 are red, so $3 + 2$, $3 + (2 \times 2)$, $3 + (2 \times 3)$, ..., $3 + (2 \times g)$..., are all red. But $3 + (2 \times g) \ge g$, so it must be green. This contradiction shows that there are no more patterns starting with a green 1. The same argument works when 1 is red. Therefore, there are 6 patterns of coloring positive integers. Note that the first green number cannot be 3 or greater, because then two smaller numbers whose sum is the first green number would also have to be green, creating a contradiction.

28 (E) 14

Let A, B, C, D, E be the five points from left to right. If one of the primitive segments, AB, BC, CD, or DE, has length greater than 9, then 22 = AE = AB + BC + CD + DE > 9 + 2 + 5 + 6 = 22, so each primitive segment has the length less than or equal to 9. If 9 is not among primitive segments, then 22 = AE = 2 + 5 + 6 + 8 = 21. This contradiction indicates that 9 must be the length of a primitive segment but 8 cannot, since 9 + 8 + 2 + 5 = 24 > 22. Therefore, the lengths of primitive segments are 2, 5, 6, 9. The segments 2 and 5 can't be adjacent since 2 + 5 = 7 is not on the list. 9 alone can't be between 2 and 5 since 11 = 2 + 9 and 14 = 5 + 9 can't be both on the list; only one of them could be k. 6 alone can't be between 2 and 5 since 11 = 5 + 6 and 13 = 5 + 6 + 2 can't be both on the list; only one of them could be k. Thus, both segments, 6 and 9, must be between 2 and 5, so 2 and 5 are at the endpoints of AE and 6 and 9 are in the middle. We may assume (changing the direction from left to right if needed) AB = 2 and DE = 5. Then BC can't be 9 since 2, 9, 6, 5 would produce two segments with the same length 11. Therefore, (assuming AB = 2) there is only one option for the primitive segments. AB = 2, BC = 6, CD = 9, and DE = 5. In the figure given below, we see the points with distances labeled.

29 (D) 5

We can make squares of area, 1, 4, and 9. Further, we can rotate our squares to make squares of area 2 and 5. Thus there are 5 total squares.

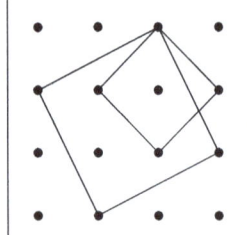

30 (D) 6

No multiplication factor can be 1 (no repetition of digits) and one factor must be 2 to keep the product below 10. The other factor is either 3 or 4 for the same reason.
If the product is 2 × 3 = 6, then the difference can be either 5 or 1 since 6 − 2 = 4, 6 − 4 = 2, and 6 − 3 = 3 are already excluded. At this stage, the digits 1, 2, 3, 5, and 6 have been used. For the next subtraction, we can only use digits 4, 7, 8, and 9. However, each of the possible subtractions, 9 − 4 = 5, 9 − 7 = 2, 9 − 8 = 1, 8 − 4 = 4, 8 − 7 = 1, and 7 − 4 = 3, either contains a duplication of digits or uses 1, 2, 3, or 5. None of the subtractions is valid. Therefore, the product must be 8, and the factors are 2 and 4 (in any order).
The subtraction under the multiplication can be 8 − 1 = 7, 8 − 7 = 1, 8 − 3 = 5, or 8 − 5 = 3. So far, we have used the digits 2, 4, 8, and two odd digits. Only one even digit, namely 6, is left for further operations. If "?" is not 6, then the addition involves two odd digits, and the sum is even. Thus, "?" must be 6. However, if "?" is 6, then the second subtraction in the diagram involves odd numbers, and the result, "?", is even. This contradiction shows that "?" must indeed be 6.

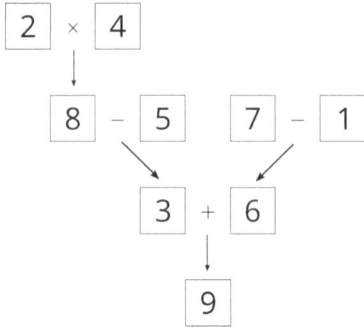

2017

3 Point Solutions

1 **(B) 10:00 a.m.**
From 5:00 p.m. to midnight is 7 hours. 10 hours later is 10:00 a.m.

---

2 **(C) 11**
Between Yana and Xena, to the left from Yana, there are 3 girls. Between Yana and Xena, to the right from Yana, there are 6 girls, so there are 1 + 1 + 3 + 6 = 11 girls in the group.

---

3 **(C) 16**
−17 − 16 = −33

---

4 **(A) $\frac{1}{2}$**
Each long bicolor strip is half white, so the total white area is $\frac{1}{2}$ of the area of this isosceles triangle.

---

5 **(B) $\frac{5}{2}$ = 2.5**
Dividing gives 5 ÷ 2 = 2.5.

---

6 **(E) 24 m**
The outer rectangle includes the lengths 2 m, 3 m, 3 m, 4 m each twice beyond the perimeter of the inner rectangle. Thus it has an extra perimeter of 2(2 + 3 + 3 + 4) = 24 m.

---

7 **(D)**
The holes are symmetric with respect to the diagonal. It cannot be (A) because we would see 4 holes. (D) works because the shorter folding segments don't make new holes.

---

8 **(D) 8**
The only different 3 positive integers summing to 7 are 1, 2, and 4, which have a product of 8.

---

9 **(B) 10 cm²**
The outer colored area is 16cm² − 9cm² and the inner colored area is 4cm² − 1cm², so the colored area is 7 + 3 = 10 cm².

---

10 **(A) 2**
Together they have 60 euros. Thus to be equal, they must each have 12 euros. Yvonne should give 2 euros to each sister.

4 Point Solutions

11 **(E) $\frac{5}{12}$**
The distance between Annie and Bob is $1 - \left(\frac{1}{3} + \frac{1}{4}\right) = \frac{5}{12}$ of the length of the pole.

---

12 **(A) $\frac{1}{2}$**
$\frac{5}{6}$ of the audience are children and $\frac{3}{5}$ of them are girls, so $\frac{5}{6} \times \frac{3}{5} = \frac{1}{2}$ of the audience are girls.

---

13 **(D) 40**
For each equilateral triangle the horizontal side is dashed and two other sides form a segment of the black path, so the length of the black segment is twice the length of the horizontal side. Therefore, the length of the black path is 2 × 20 = 40.

---

14 **(A) 14**
Let E, I, R, and Z represent the ages of Ema, Iva, Rita, and Zina, respectively. We have the following information: ($Z + E$) and ($Z + R$) are multiples of 5. ($R - E$) is a positive integer and a multiple of 5, with $E < R$.
Among the six possible differences (14 − 12, 14 − 8, 14 − 3, 12 − 8, 12 − 3, and 8 − 3), only 8 − 3 = 5 is a multiple of 5. Hence, Rita is 8 and Ema is 3. ($Z + 3$) and ($Z + 8$) are multiples of 5, so $Z = 12$. Therefore, Iva is 14.

15 (E) 840
252 = men − women, which is (1 − .35) − .35 = .3 of the total number of runners.
Thus 252 is $\frac{3}{10}$ of the total, so the total number of runners was $252 \div \frac{3}{10} = 252 \times \frac{10}{3} = 840$.

---

16 (A) 63
35 − 22 is equal to the sum of all five numbers minus the sum of the numbers in the first three boxes, which is also equal to the sum of the numbers in the last two boxes. Therefore, the sum of the numbers in the last two boxes is 13. We know that one of the numbers is 4, so the number in the second gray box is 13 − 4 = 9.
Similarly, 10 = 35 − 25 is equal to the sum of the numbers in the first two boxes. We know that 3 is one of these numbers, so 10 − 3 = 7 is the number in the first gray box. The product of the numbers in the gray boxes is 7 × 9 = 63, which indeed satisfies all the requirements. We can have 3, 7, 12, 9, and 4.

---

17 (B) 16
Simon makes 9 − 1 = 8 cutting marks and Barbara makes 8 − 1 = 7 cutting marks. None of the cutting marks will overlap because 8 and 7 are relatively prime. There is a total of 15 cutting marks. Each cut splits a piece into two pieces. Therefore making 15 cuts gives 16 pieces.

---

18 (B) 4
Let P be the intersection point of the two segments connecting points on the opposite sides. Let h_1 be the distance from P to the bottom side, and h_2 be the distance from P to the top side. Then $h_1 + h_2 = 8$, and the shaded area of both triangles is:

$\frac{1}{2} \cdot 1 \cdot h_1 + \frac{1}{2} \cdot 1 \cdot h_2 = \frac{1}{2} \cdot 1 \cdot (h_1 + h_2) = \frac{1}{2} \cdot 8 \, cm^2 = 4 \, cm^2$.

---

19 (B) 14
We have to select two days from the list of seven days. It can be done in $\frac{7 \cdot 6}{2} = 21$ ways, but some pairs are not valid for jogging schedules. The invalid pairs are {Mo, Tu}, {Tu, We}, {We, Th}, {Th, Fr}, {Fr, Sa}, {Sa, Su}, and {Su, Mo}. Therefore, the number of possible schedules is 21 − 7 = 14.

---

20 (D) 22
Let us call the top middle X, and the middle square C. Then $X + C = 3 + C$, so $X = 3$. The sum of the numbers in any two cells with a common edge is 5, and we can fill in all squares. The sum of all numbers is $5 \times 2 + 4 \times 3 = 22$.

2	3	2
3	2	3
2	3	2

5 Point Solutions

21 **(C) 91°**
Let $\alpha < \beta < \gamma$ be the three angles in a triangle. Then $\alpha + \beta + \gamma = 180°$, and $180° = \alpha + \beta + \gamma > \alpha + \beta + \beta > 2\beta$, so $\beta < 90°$. Since α and β are integers, we have $2° \leq \beta \leq 89°$.
The sum of the smallest and largest angles is $\alpha + \gamma$, and $\alpha + \gamma = 180° - \beta$. We know that $2° \leq \beta \leq 89°$, so $91° \leq (180° - \beta) \leq 178°$.
If we consider the case where $\alpha = 1°$, $\beta = 89°$, and $\gamma = 90°$, then $\alpha + \gamma = 91°$. Therefore, 91° is the minimal sum of the smallest and largest angles in a triangle.

---

22 **(C) 18**
R stands for each kangaroo looking to the right and L stands for each kangaroo looking to the left, so RRRLLRRRLL is the initial configuration. Let's focus on the kangaroos facing left. After 3 jumps of the first L to the left, the configuration changes to LRRRLRRRLL. The next L is also making 3 jumps to the left, so the current configuration is LLRRRRRRLL. Each of the next two L's is making 6 jumps to the left, so the final configuration is LLLLRRRRRR and no more jumps are possible. The total is $3 + 3 + 2 \times 6 = 18$.

---

23 **(B) 6**
Here is a possible distribution of the results: $1 + 5$, $2 + 5$, $3 + 5$, $4 + 2$, $5 + 2$, $6 + 2$, $7 + 2$, $8 + 2$, and $9 + 2$. The results are 6, 7, 8, 6, 7, 8, 9, 10, and 11. There are 6 different results. To show that 6 is the smallest number of different results, let's consider the following six numbers: 1, 2, 3, 7, 8, and 9. If we add 2 to each of them, we get 3, 4, 5, 9, 10, and 11. If we add 5 to each of them, we get 6, 7, 8, 12, 13, and 14. All of these 12 results are different numbers. Therefore, by adding randomly either 2 or 5 to each of the original numbers, we will always produce at least 6 different results.

---

24 **(A) 8**
Let B_k (for $k = 1, 2, 3, \ldots, 20$) be the bus that left the airport $3k$ minutes before the car left the airport. The bus B_k will reach the center in $(60 - 3k)$ minutes, and the car will reach the center in 35 minutes. Therefore, the car will pass the kth bus if $(60 - 3k) > 35$. We can calculate the times for the buses $B_1, B_2, B_3, \ldots, B_9$ to reach the center:
$B_1 : 60 - 3(1) = 57$ minutes
$B_2 : 60 - 3(2) = 54$ minutes
$B_3 : 60 - 3(3) = 51$ minutes
$B_4 : 60 - 3(4) = 48$ minutes
$B_5 : 60 - 3(5) = 45$ minutes
$B_6 : 60 - 3(6) = 42$ minutes
$B_7 : 60 - 3(7) = 39$ minutes
$B_8 : 60 - 3(8) = 36$ minutes
$B_9 : 60 - 3(9) = 33$ minutes
Out of these buses, the car will pass 8 buses, since $(60 - 3k) > 35$ for $k = 1, 2, 3, 4, 5, 6, 7, 8$, but $(60 - 3 \cdot 9) = 33 \leq 35$. Therefore, the car will pass 8 buses.

---

25 **(D) 32**
The total area is $5 \times 5 = 25$. There are 16 square units along the edge, and half of this edge area is black, so it makes up $\frac{8}{25} = 32\%$ of the total area.

---

26 **(A) 2**
The given sequence is 2, 3, 6, 8, 8, 4, 2, 8, 6, 8, 8, 4, 2, 8, 6, 8, 8, 4, 2, 8, 6, 8, 8, 4, 2, 8, ...
Observing the pattern, we notice that after the first two digits (2, 3), the sequence 6, 8, 8, 4, 2, 8 repeats indefinitely.
We can observe that every digit in the (7 + a multiple of 6)th position is equal to 2. For example, the 7th digit, 13th digit, 19th digit, 25th digit, and so on, are all equal to 2. To find the digit in the 2017th position, we can write 2017 as $7 + 6 \times 335$. Therefore, the 2017th digit will be the same as the 7th digit, which is 2.
Hence, the 2017th digit in the sequence is 2.

27 (D) 39

The top right vertex of the front face is the reference point. We can move from it in three directions. The depth direction (d-direction) is the direction perpendicular to the front face, the height direction (h-direction) is the direction perpendicular to the top face, and the width direction (w-direction) is the direction perpendicular to the right face. There are 5 layers in each direction. In each direction, we enumerate them as 1, 2, 3, 4, 5 starting from the reference point. Each unit cube of the big cube belongs to one d-layer, one h-layer, and one w-layer, so the unit cube can be described by three numbers as (d, h, w), where $1 \le d, h, w \le 5$.

There are three tunnels perpendicular to the front face. Their unit cubes are: (1, 2, 2), (2, 2, 2), (3, 2, 2), (4, 2, 2), (5, 2, 2); (1, 3, 3), (2, 3, 3), (3, 3, 3), (4, 3, 3), (5, 3, 3); and (1, 4, 4), (2, 4, 4), (3, 4, 4), (4, 4, 4), (5, 4, 4).

There are three tunnels perpendicular to the top face and their unit cubes are: (2, 1, 2), (2, 2, 2), (2, 3, 2), (2, 4, 2), (2, 5, 2); (3, 1, 3), (3, 2, 3), (3, 3, 3), (3, 4, 3), (3, 5, 3); and (4, 1, 4), (4, 2, 4), (4, 3, 4), (4, 4, 4), (4, 5, 4).

There are three tunnels perpendicular to the right face and their unit cubes are: (2, 2, 1), (2, 2, 2), (2, 2, 3), (2, 2, 4), (2, 2, 5); (3, 3, 1), (3, 3, 2), (3, 3, 3), (3, 3, 4), (3, 3, 5); and (4, 4, 1), (4, 4, 2), (4, 4, 3), (4, 4, 4), (4, 4, 5).

Together there are 9 tunnels; they contain 9×4 unit cubes that are not shared and 3 unit cubes, (2, 2, 2), (3, 3, 3), and (4, 4, 4), that are shared.

Hence, $9 \times 4 + 3 = 39$ is the number of unused cubes.

28 (E) 320

The second runner is slower than the first one. The speed of the second runner is $\frac{720}{5}$ m/min, and the speed of the first runner is $\frac{720}{4}$ m/min. The proportion of the speeds (second:first) is 4:5. Therefore, the answer is $\frac{4}{4+5} \times 720 = 4 \times 80 = 320$.

29 (D) 10

The top two rows cannot have more than 2 odd numbers. Similarly, the top three rows cannot have more than 4 odd numbers. Indeed, if the third row consists of only Os (odd numbers), then all numbers above are Es (even numbers), so there are only 3 odd numbers. If the third row contains only 2 (or fewer) Os, then the total of Os is less than or equal to (2 + 2) = 4.

The top four rows cannot have more than 7 odd numbers. This is true when the fourth row has at most 3 Os since 3 + 4 = 7. When the fourth row consists of four Os, then the top four rows have only 4 Os. If the whole pyramid of five rows has only 3 Os in the bottom row, then the numbers of Os in this pyramid does not exceed 3 + 7 = 10 Os.

If the bottom row consists of 5 Os, then the whole pyramid has exactly 5 Os. Thus, we have to analyze only pyramids with 4 Os in the bottom row. By placing O O E O O in the bottom row, we can have 10 O's. Note that O E O O E in the bottom row also works.

30 (D) $\frac{1}{12}S$

Let h be the distance between AB and DC. Then, Area of $\triangle AED$ + Area of $\triangle DME = \frac{1}{2}h \cdot DM$ and Area of $\triangle BFC$ + Area of $\triangle MCF = \frac{1}{2}h \cdot MC$, so Area of $\triangle AED$ + Area of $\triangle BFC$ + Area of $\triangle DME$ + Area of $\triangle MCF = \frac{1}{2}h \cdot DM + \frac{1}{2}h \cdot MC = \frac{1}{2}h \cdot (DM + MC) = \frac{1}{2}h \cdot DC = \frac{1}{2}S$. As stated, Area of $\triangle AED$ + Area of $\triangle BFC = \frac{1}{3}S$, so Area of $\triangle DME$ + Area of $\triangle MCF = \frac{1}{2}S - \frac{1}{3}S = \frac{1}{6}S$. Area of $\triangle DME$ + Area of $\triangle MCF$ + Area of the quadrilateral $EOFM$ = Area of $\triangle CDO = \frac{1}{4}S$, so the area of the quadrilateral $EOFM = \frac{1}{4}S - \frac{1}{6}S = \frac{1}{12}S$.

2019

3 Point Solutions

1 (E)

Any number that ends in any of the even digits (0, 2, 4, 6, or 8) is even. (E) has all even numbers.

---

2 **(E) 2 and a half**
One quarter of an hour is 15 minutes. Ten quarters is 150 minutes, or two and a half hours.

---

3 **(C) 20**
This way of removing the cubes removes the middle cube from each of the six faces and the cube in the middle of the large cube. 27 − 7 = 20 cubes remain.

---

4 (D)

In the given figure the white ring passes through the gray. This excludes answers (B), (C), and (E). Also, in the given figure the black and gray rings are free of each other. This excludes answer (A). One can check that answer (D) is correct (white ring through the other two, gray and black independent).

---

5 (D)

Try starting the drawing from a point where multiple segments meet. All the diagrams except (D) can be drawn.

---

6 **(D) 40**
Each of the 5 friends gave 4 cupcakes to other friends, so 5 × 4 = 20 cupcakes were given and eaten. The five friends had 20 × 2 = 40 cupcakes at the start.

---

7 **(A) Victor**
If A < B denotes "A is slower than B," we are given Manfred < Lotar, Victor < Jan, Jan < Manfred, and Victor < Eddy. This can be written as V < J < M < L. Since V < E, this indicates that V (Victor) is the slowest and finished last.

---

8 **(B) 58**
Since there are six pages that end with 8 and five pages that end with 0, the first page (which has the smallest number) would be 8. Consequently, the next larger number you can create the way described is 10. Logically, the next pages would be 18, 20, 28, 30, 38, 40, 48, 50, and the last page, 58.

---

9 (D) $\frac{4}{9}$
The large square is divided into four sections, with one of those four sections being divided into nine smaller sections, of which seven are colored. Notice that in the big square we have 4 × 9 = 36 small squares. This means that $\frac{1}{4} + \frac{7}{36} = \frac{4}{9}$ of the square is colored.

---

10 **(A) 60**
Since both boys have the same number of apples, we can imagine converting Andrew's six piles into five piles. We know from he problem that each of the five piles will need two more apples, and these have to come from the sixth pile. So, the sixth pile is divided into five sets of two apples each, meaning it had 10 apples to start. Since Andrew had six such piles, he had 60 apples.

4 Point Solutions

11 **(A) 5, 6, and 7**

```
  1234
  21?7
 +??26
 10126
```

The three numbers turn out to be 1243, 21**5**7, and **67**26.

---

12 **(B) 60°**
Triangle PQR is isosceles, so ∠PRQ = 20° and ∠PQR = 140°. Triangle PQS is also isosceles, thus ∠PQS = ∠QSP, and they are equal to (180° − 20°) ÷ 2 = 80°. Consequently, ∠QSR = 180° − 80° = 100°, so ∠RQS = 180° − 100° − 20° = 60°.

---

13 **(E)**
Find a black corner square and use it to place the L-shaped piece. Square (E) has two black corner squares on the same side, but the ends of the L-piece are white, so we cannot place the L-piece. Therefore, square (E) cannot be formed. Here is how we can form the other squares:

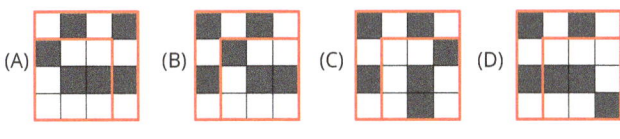

---

14 **(B) 2**
Dora shook hands with everyone, so Alan finished his handshakes with her and had no more left. Claire shook hands with 3 people but not with Alan since he had no more handshakes available. Therefore, Claire shook hands with Bella, Dora, and Erik. Bella finished her 2 handshakes, one with Dora and one with Claire. By counting, we can see that Erik shook hands twice.

---

15 **(C) 3**
For the first 20 shots, Jane scored $20 \times \left(\frac{55}{100}\right) = 11$ times. Five shots later, after 25 shots, she scored $25 \times \left(\frac{56}{100}\right) = 14$ times. This means she scored an additional 14 − 11 = 3 shots in the last 5 shots.

---

16 **(C) 5**
The figure below shows the pieces after the paper is cut and unfolded. The dashed lines represent the folds. There are 5 squares.

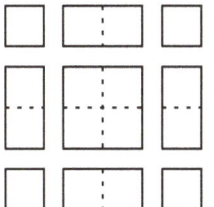

---

17 **(D) 7**
From the data it easily follows that there are 3 dogs. 18 are not cows and 16 are not cats. In other words, dogs, cats, and kangaroos (i.e., not cows) total to 18 so, without the 3 dogs, there are 15 cats and kangaroos. Also, dogs, cows and kangaroos (i.e., not cats) total 16, so there are 13 cows and kangaroos. Solving, we find that there are 7 kangaroos, 8 cats, and 6 cows.

---

18 **(B) 12 cm²**
Since there are 5 rectangles on the base, the length of each rectangle is 10 ÷ 5 = 2 cm. Similarly, the height is 6 ÷ 4 = 1.5 cm. The shaded area is equal to 14 rectangles minus the triangle, so it is $(14 \times 2 \times 1.5) - (10 \times 6 \times \frac{1}{2}) = 12$ cm².

19 **(C) 5:4**
Let the heights of the candles be h and k, respectively. Since the first candle burns in 6 hours, in 3 hours its height will be $\frac{h}{2}$. Similarly, 3 hours later, the (remaining) height of the second candle will be $\frac{5k}{8}$. We are given $\frac{h}{2} = \frac{5k}{8}$, so $\frac{h}{k} = \frac{5}{4}$.

---

20 **(C) 16**
The solution is unique, as shown. You start building it by continuing "up" from the given match as you cannot go "straight" nor "down" because of the 0 at a nearby square. You can also straight away put the 3 matches around each of the 3's (the empty border is the one that is also a border to a square with a 0). You can finish the path from there.

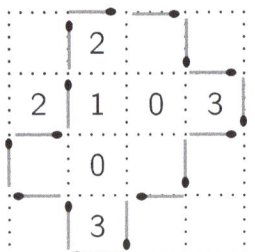

5 Point Solutions

21 **(B) 32**
On one of the two sides of 7 the numbers are consecutive up to 23. This means they are the 15 numbers 8 to 22, inclusive. By symmetry we have another 15 numbers to the other side of 7, making a total of 15 + 15 + 2 = 32 numbers in the circle.

---

22 **(B) $75**
Since Liam had $50 at the beginning, it means he had $50 + $10 = $60 after selling 40 bottles. So, he sold each for $60 ÷ 40 = $1.50. Selling all 50 bottles for $1.50 each amounts to $50 × 1.50 = $75.

---

23 **(C) 5**
Consider the five shaded squares on the left figure. They have no common border, and each must contain a green stick. So, there are at least 5 green sticks. The figure on the right shows a situation as described in the problem, where the green sticks are exactly 5; this is the smallest possible number.

 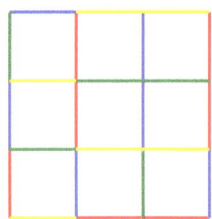

24 (E)

The red edges show where the line is not continous after nets (A), (B), (C), (D) are assembled. In the case of (E) (not shown), it is continuous.

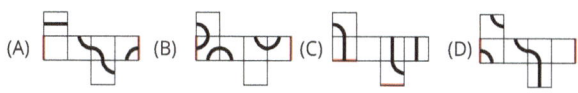

25 (E) 6

The first day Elisabeta ate $\frac{1}{10}$ of the 60 chocolates, so she left $\left(\frac{9}{10}\right) \times 60$ for the next day. The day after that, she left $\left(\frac{8}{9}\right) \times \left(\frac{9}{10}\right) \times 60$ chocolates, and so on up to

$\left(\frac{1}{2}\right) \times \left(\frac{2}{3}\right) \times \left(\frac{3}{4}\right) \times \left(\frac{4}{5}\right) \times \left(\frac{5}{6}\right) \times \left(\frac{6}{7}\right) \times \left(\frac{7}{8}\right)$
$\times \left(\frac{8}{9}\right) \times \left(\frac{9}{10}\right) \times 60$

When carrying out the calculation, we save time if we observe that the product is telescopic (cancelations all along between a numerator and a denominator at all consecutive fractions). The final answer is

$\left(\frac{1}{10}\right) \times 60 = 6$

26 (A) 5 and 8

Circles 2 and 6 are neighbors so they must have different colors. Circles 5 and 8 are both neighbors to 2 and 6, so they must have the same color (namely the third color, not that appearing in 2 or 6). This shows that answer (A) is correct. The following is a counter-example to each of the other answers (B)–(E).

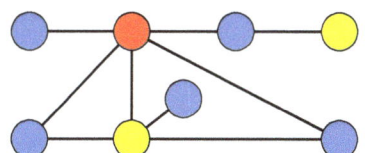

27 (C) 250

The amounts Flora and Ria have are $5N$ and $3N$, respectively. After Ria spent $160, the ratio of their money is given as $\frac{5N-160}{3N} = \frac{3}{5}$. Solving this equation, we find that $5N = 50$. Substituting $N = 50$ back into the expressions, we can calculate that Ria had $5 \times 50 = \$250$.

28 (E) 7

Suppose there are n teams. Each player plays with 3 players from each of the other $n - 1$ teams, so they play $3(n - 1)$ games. This means that the total number of games is $\frac{1}{2} \times 3n \times 3(n - 1)$ (notice we divide by 2 because each game is counted twice). This sets up the inequality $\frac{9}{2} n(n - 1) \le 250$.
The function $n(n - 1)$ is increasing, so we can find a suitable value for n by trying a few values. For example, if we plug in $n = 8$, we would get 252 games, which is too large. However, the next lowest value, $n = 7$, gives 189 games, which is less than 250.

29 (E) $\frac{3}{8}$

From the equality of the triangles △ABQ and △ABP, it follows that ∠QAB = ∠PBA, and so triangle SAB is isosceles. Also, △ARB is isosceles, from where it follows that R and S are on the axis of symmetry RT of the square. As T is the midpoint of AB, it follows that $ST = \frac{1}{2}BQ = \frac{1}{4}a$, where a is the side of the square, so $RS = \frac{3}{4}a$. It follows that the shaded area equals
$2 \times \frac{1}{2} \times RS \times AT = \frac{3}{4}a \times \frac{1}{2}a = \frac{3}{8}a^2 = \frac{3}{8}(ABCD)$

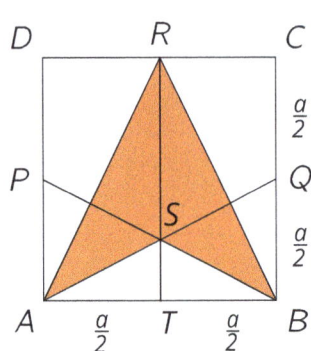

30 (D) 96

Let the number of passengers in the first five carriages be P, Q, R, S, T, and let X be the number of passengers in the sixth carriage. We are given that $P + Q + R + S + T = 199 = Q + R + S + T + X$. It follows that $P = X$. In a similar way, we see that the number of passengers in the carriages is a periodic sequence, as shown in the figure. The two middle carriages have S and T passengers, respectively, so the problem asks for $S + T$. As all passengers amount to 700, we conclude that the last three carriages satisfy $P + Q + R = 700 - 3 \times 199 = 103$. So, $S + T = (P + Q + R + S + T) - (P + Q + R) = 199 - 103 - 96$.

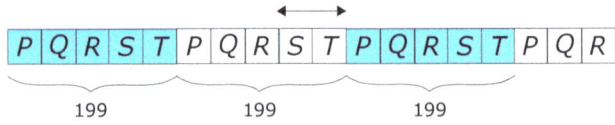

2021

2021

3 Point Solutions

1 (A) Sagittarius

Reflecting the Sagittarius symbol along the diagonal line from the bottom left to the top right does not change the symbol, so it is an axis of symmetry.

---

2 (E) 50%

If we rotate every shaded section on the right side of the disk by 180°, notice that the left side of the disk is completely shaded and the right side is completely white. Additionally, rotation of a section does not change its area. Thus, 1/2, or 50%, of the figure is shaded.

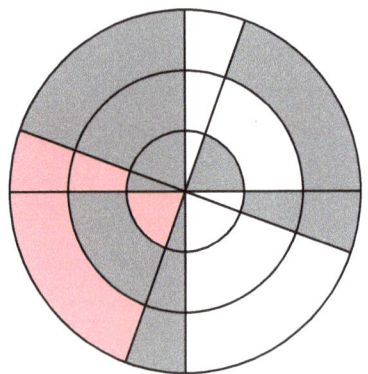

---

3 (D) 84

$$\frac{20 \cdot 21}{2+0+2+1} = \frac{20 \cdot 21}{5} = 4 \cdot 21 = 84$$

---

4 (B) 6

The only four-digit numbers whose digits, when read from left to right, are consecutive integers in ascending order, are: 1234, 2345, 3456, 4567, 5678, and 6789.

---

5 (A) −100

Label the pieces in the order shown a, b, c, d, and e. Since piece a's right side is a vertical line, it must be the right-most piece. Similarly, since piece c's left side is a vertical line, it must be the left-most piece. In order to find the second piece from the left, note that its left side must fit into c's right side. The only two pieces whose left sides can fit into c's right side are a and e. But a is the right-most piece, so e must be the second piece from the left. In order to find the second piece from the right, note that its right side must fit into a's left side. The only two pieces whose right sides can fit into a's left side are b and c. But c is the left-most piece, so b must be the second piece from the right. Now that we know that c is the left-most piece, e is the second piece from the left, b is the second piece from the right, and a is the right-most piece, this leaves d as the middle piece. We can now read the pieces in order: c, e, d, b, $a \Rightarrow 2 - 102 = -100$.

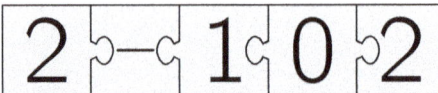

---

6 (A)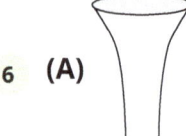

Each vase is filled to half its capacity. The pink plane shown in each illustration below intersects each vase halfway between the base and the top. Notice that vases B, C, and E are symmetrical with respect to this plane, so the water level will be exactly in the middle. In Vase D, the water level will be below the middle plane, since the lower part of the vase is wider than the higher part. In Vase A, the water level will be above the middle plane, since the lower part is narrower than the higher part.

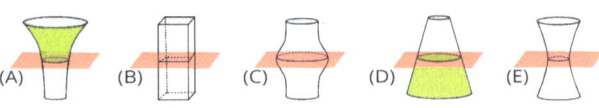

7 (B) 13837

$\overline{AB} + \overline{CD} = 137$, so $\overline{ADCB} + \overline{CBAD} =$
$1000A + 100D + 10C + B$
$\quad + 1000C + 100B + 10A + D$
$\quad\quad = 1010A + 101B + 1010C + 101D$
$\quad\quad\quad$ (combine like terms)
$\quad\quad = 101 \cdot ((10A + B) + (10C + D))$
$\quad\quad = 101 \cdot (\overline{AB} + \overline{CD})$
$\quad\quad = 101 \cdot 137 = 13837$

8 (E)

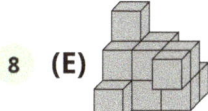

Consider the face that is facing us, out of the page. The lower right-hand corner of this face is white, not gray, so this eliminates choice (A). Now consider the face that is opposite us, facing into the page. The middle right-hand edge piece of this face (relative to our perspective) is black, not gray, so this eliminates choice (B). And the lower right-hand corner of this face is gray, so this eliminates choice (C). Finally, notice that the middle left-hand edge of the top face is black, so this eliminates choice (D). The only remaining choice is (E), which one can verify contains all of the gray cubes and no non-gray cubes.

9 (B)

Since there are 10 possible digits, a 180° rotation corresponds to adding or subtracting 5. The original digits are 6, 3, 4, and 8, and in choice (B), the digits are 1, 8, 9, and 3. $|1 - 6| = |8 - 3| = |9 - 4| = |3 - 8| = 5$, therefore choice (B) is correct.

10 (E) Aaron is 30 cm shorter than Erin.

Let Aaron's height be a, Byron's height be b, Caron's height be c, Darren's height be d, and Erin's height be e. We are given: $b = a + 5$, $b = c - 10$, $d = c + 10$, $d = e - 5$. $e - a = (d + 5) - a = (c + 15) - a = (b + 25) - a = (a + 30) - a = 30$. Therefore, Aaron is 30 cm shorter than Erin.

4 Point Solutions

11 (D) 45

Since Neil breaks off a piece of chocolate with a height of 2 to obtain 12 squares, the width of this piece must be 6 squares. Since $9 \ne 6$, Jack's strip of chocolate must not be parallel to Neil's. Therefore Jack's strip has height 9 and width 1. As shown in the illustration, the original dimensions of the chocolate bar must have been 6×11, and the remaining bar must have dimensions 5×9 for a total of 45 squares left. The illustration shows Neil's piece in blue and Jack's piece in red.

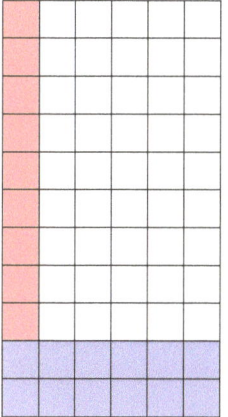

12 (E) 500 g

We can find the weight of three-fifths of a jar of water by subtracting one weight from another. 740 grams − 560 grams = 180 grams. Therefore, the weight of one-fifth of a jar of water is 180 grams ÷ 3 = 60 grams. Therefore, the jar itself weighs 560 grams − 60 grams = 500 grams.

13 (C) 4 cm²

The two dashed lines indicate lines of symmetry of the diagram. As shown, the width of each yellow square (and the base of each black triangle) must be 1, and the height of each black triangle must also be 1. Since there are 8 black triangles, each with base 1 and height 1, their total area is $8 \cdot (\frac{1}{2} \cdot 1 \cdot 1) = 4$ cm².

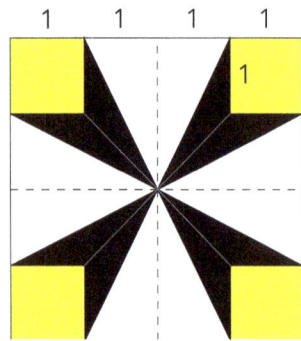

14 (B) 2.5

There are two rows of wooden planks. The top row consists of 13 planks and the bottom row consists of 12 planks. The ends of each bottom plank are covered by top planks. Cut off these $12 \cdot 2 = 24$ ends, so that the top planks and the shorter bottom planks cover a distance of 6.9 meters without any gaps or overlaps. Now, extend this 6.9-meter row by 24 ends (the overlaps) to see a long row with length $25 \cdot 30$ cm, since its length is the same as the combined length of 25 non-overlapping planks of wood. The combined length of 24 overlaps is $(25 \cdot 30 - 690) = 60$ cm, so the length of one overlap is $60 \div 24 = 2.5$ cm.

15 (D) 20

The sum of the angles at the center of a star must equal 360°, because when put together they go all the way around the center point and back to where they started. Therefore, in the star shown in the diagram, each central angle measures $360° \div 5 = 72°$.

Since the larger acute angle in each right triangle is 72°, the smaller angle must be $90° - 72° = 18°$. Now, since the new star's central angles must also sum to 360°, there must be $360° \div 18° = 20$ triangles in the new star.

16 (C) 4

Call the side length of the smallest unknown square x, as shown in the diagram. The side length of the square at the upper left corner is $x + 1$ and the diagram shows the side lengths of the other squares. If you move from the left-most edge of the diagram to the right-most edge along the top edges, then the total distance covered in the horizontal direction is $x + 1 + x + h$. If you do the same along the bottom edges, then the distance is $x + 2 + x + 3$, so $x + 1 + x + h = x + 2 + x + 3$. This simplifies to $h = 4$.

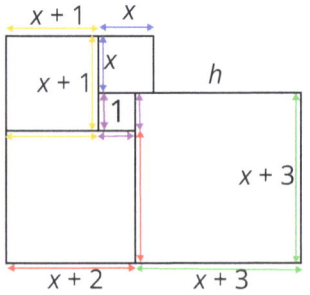

17 (B) 1

Suppose that Eric had gotten every single question on the quiz correct, without leaving any questions blank. Then, he would have gotten $7 \cdot 20 = 140$ points. However, he actually got 100 points. This means that he lost 40 points by leaving some questions blank and getting some questions wrong.

Let b be the number of questions Eric left blank, and let w be the number of questions Eric got wrong. We know that b and w are nonnegative integers. Now, note that for every question that Eric left blank, he got 7 fewer points than he would have gotten if he had answered it correctly. And for every question he got wrong, he got 11 fewer points than he would have gotten if he had gotten it correct. Therefore, $7b + 11w = 40$. Since $b \geq 0$, $11w \leq 7b + 11w = 40$, or $11w \leq 40$. Since w is an integer, $w \leq 3$ (in other words, if $w \geq 4$, then $11w > 40$).

We now know that w must be 0, 1, 2, or 3. Testing these cases, we get that $w = 3$ and $b = 1$ is the only solution.

18 **(C) 6**

Label the points as shown in the image. Since the fold is at a 45° angle, $\angle ABF = 90°$ and $ABFG$ is a rectangle. Therefore, $\overline{BF} = \overline{AG} = 4$. Additionally, since $\triangle BCF$ is a 45–45–90 isosceles triangle, $\overline{CF} = \overline{BF} = 4$. Also, since the rectangle is folded, we know that $\overline{AB} + \overline{BF} + \overline{FE} = 13$. Now, $P = \overline{AG} \cdot \overline{AB} = 4x$ and $Q = \overline{CF} \cdot \overline{FE} = 4 \cdot \overline{FE}$. Therefore, $4x = P = 2Q = 8\overline{FE}$, or $x = 2\overline{FE}$. $\overline{AB} + \overline{BF} + \overline{FE} = 13$ can now be simplified to $x + 4 + \frac{x}{2} = 13$. $\frac{3x}{2} = 9$, or $x = 6$.

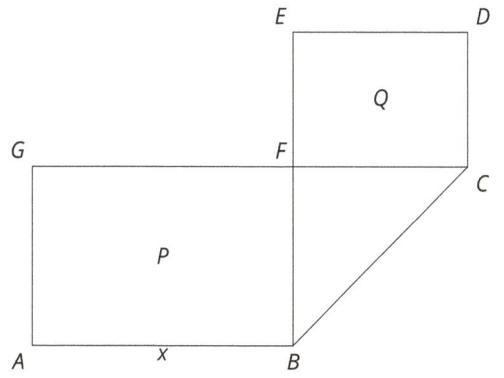

---

19 **(E) Christy took as many pears as Lily got apples.**

Let p be the number of pears in the box. Clearly, there are $2p$ apples in the box. Let a be the number of apples that Lily took. Since all of the apples were taken by either Christy or Lily, this means that Christy took $2p - a$ apples. Since Christy got twice the number of fruits as Lily and the only remaining fruits are pears, we can calculate the number of pears that Christy took. Let this number be x. By similar reasoning to the apples, we know that Lily took $p - x$ pears.

$(2p - a)$ apples + x pears =
$= 2 \cdot (a$ apples + $(p - x)$ pears)
$2p - a - x = 2p + 2a - 2x$
$-a - x = 2a - 2x$
$-a + x = 2a$
$x = a$

Since Lily took a apples and Christy took x pears, we know that Christy took as many pears as Lily got apples.

---

20 **(C) 3 km**

Let m be the distance between Uphill and Downend, d be the distance between Uphill and Middleton, and u be the distance between Downend and Middleton. Adding together the equations that we are given:

$u + d - m = 1$ (1)
$m + d - u = 5$ (2)
$m + u - d = 7$ (3)
$\overline{}$
$u + d + m = 13$ (4)

Now, subtracting (1) from (4), we get that $2m = 12$, or $m = 6$. Performing similar computations for u and d, we get that $u = 4$ and $d = 3$. Therefore, the shortest direct path is d, which has length 3.

5 Point Solutions

21 **(C) 30%**

Let the integers n and d be the numerator and denominator of the original fraction. We want to find some $p > 0$ such that $\frac{n + 0.4n}{d - pd} = \frac{2n}{d}$. Or, $\frac{1.4n}{(1-p)d} = \frac{2n}{d}$.

Multiplying both sides by $\frac{d}{n}$, we get $\frac{1.4}{(1-p)} = 2$ or $1 - p = 0.7$. Therefore, $p = 0.3 = 30\%$.

---

22 **(D) D**

Let the face corresponding to the left-most diagram be face 1, the one corresponding to the middle diagram be face 2, and the one corresponding to the right-most diagram face 3. In a tetrahedron, any set of three faces must be adjacent to the fourth face, and each pair of faces within the set must be adjacent. Indeed, we can tell that face 2 is on the left of face 1, since face 1's left edge reads DCBD from the top down, and face 2's right edge also reads DCBD. Similarly, face 3 must be on the right of face 1. Now that we know how the faces align, we can count the number of D's. The top-most D is common to face 1, 2, and 3. Then, there is a D on the right edge of face 1 (the left edge of face 3). Finally, there is a D on the left edge of face 1 (aka the right edge of face 3). Since there are only 3 D's on the faces shown and there are 4 cannon balls with each label, we know that the hidden cannon ball must be labeled D.

23 (B) 27

Let $x = \overline{ABCDE}$. We are given that $3 \cdot (200{,}000 + x) = 10x + 2$. Or, $7x = 599{,}998$. $\overline{ABCDE} = x = 85{,}714$, so $2\overline{ABCDE} = 285{,}714$. The sum of the digits is 27.

24 (B) 29

Let the number of red counters be r, and define g, b, and y similarly for green, blue, and yellow, respectively. Since the box contains only green, red, blue, and yellow, the maximum possible number of counters we could pick before we get a red counter is $g + b + y$. Only after we exhaust all of the green, blue, and yellow counters, only red counters are left and we are guaranteed a red one on our next pick. In other words, $g + b + y \leq 25 - 1$. Similarly, $r + b + y \leq 27 - 1$, $r + g + y \leq 22 - 1$, and $r + g + b \leq 17 - 1$. Adding these together, $r + g + b + y \leq 29$. Before we can conclude, we know that 29 is an upper bound for $r + g + b + y$, but we must first verify that it is possible to achieve this theoretical maximum. Turning all of the ≤ signs into = signs, we can solve the resulting system of equations to get $r = 5$, $g = 3$, $b = 8$, and $y = 13$ as a solution that satisfies all of the inequalities. Therefore, 29 is the largest number of counters that could be in the box.

25 (D) 20

There are 12 pentagons in the ball: 1 on top, 1 on the bottom, and 2 horizontal rings of 5 in the middle. Each pentagon has 5 sides, and is only adjacent to hexagons, so there are a total of $12 \cdot 5 = 60$ hexagons adjacent to pentagons. However, each hexagon is adjacent to exactly 3 pentagons, which means that each hexagon was counted 3 times (1 time for each pentagon that it borders). When we account for this overcounting, the total number of hexagons is $60 \div 3 = 20$.

Another interesting way to solve this problem is to consider the icosahedron and dodecahedron. The hexagons form the 20 faces of an icosahedron, with the pentagons being the 12 vertices. Or, the pentagons could form the 12 faces of a dodecahedron, with the hexagons forming the 20 vertices.

26 (B) 20

Divide the numbers from 1 to 2021 into three groups based on their remainders when divided by three. There are only 3 possible remainders when a number is divided by 3: 0, 1, or 2. Call these groups *residue classes*. Let the residue class r_n be the residue class whose remainder when divided by 3 is n. In other words, the three residue classes are r_0, r_1, and r_2. Now, we will show that there is a one-to-one correspondence between residue classes and the color of the kangaroos.

Since Kangaroos 1, 2, and 3 must have different colors, and Kangaroos 2, 3, and 4 must have different colors, neither Kangaroo 1 nor Kangaroo 4 can be the color of either Kangaroo 2 or Kangaroo 3. But since there are only 3 colors (red, gray, and blue), this means that Kangaroo 1 must be the same color as Kangaroo 4. This analysis can be repeated to show that Kangaroos 2 and 5 must be the same color, as must 3 and 6, 4 and 7, 5 and 8, 6 and 9, etc. Since the colors repeat every 3 kangaroos and residue classes also repeat every 3 numbers, each residue class has a unique color. Since the remainders of 2, 20, and 2021 when divided by 3 are all 2, they all belong to r_2. Bruce's guesses for the color of r_2 are gray, blue, and gray. Since only one of these guesses can be wrong, blue must be the incorrect guess. This corresponds to kangaroo 20.

27 (C) 10

Consider the moment when a termite eats through a wall. At this moment, it exits one cube and enters another cube. When the termite starts out, it immediately passes through one cube – the starting cube. After that, each time it eats through a wall, it enters another cube. Since the diagonal does not intersect any edges, the total number of walls that the termite passes through is the sum of the number of walls in each direction – x, y, and z. We can see that there are 4 walls in the x direction (parallel to the yz plane), 2 walls in the y direction, and 3 walls in the z direction. So, the total number of cubes that the termite passes through is $1 + 4 + 3 + 2 = 10$.

28 (D) 995

There are 3 possibilities for a pair: it can either be knight-knight, knight-knave, or knave-knave. In a knight-knight pair, each knight would correctly describe the other as a knight. In a knight-knave pair, the knight would correctly describe the knave as a knave, and the knave would incorrectly describe the knight as a knave. And in a knave-knave pair, each knave would incorrectly describe the other as a knight. Therefore, every two people who were described as knaves corresponds to one knight-knave pair. Since 20 people were called knaves, there must have been exactly 10 knight-knave pairs. Since the wizard divided 2020 out of the 2021 people into pairs, there must be 1 person left out. We can now determine whether that person was a knight or a knave. Suppose that the number of knights that were included is k. Each knight-knight pair adds 2 to k, each knight-knave pair adds 1 to k, and each knave-knave pair adds 0 to k. This means that if the number of knight-knave pairs is even, **then k must be even**, since knight-knight pairs and knave-knave pairs both add an even number to k. And indeed, we have just shown that there were 10 knight-knave pairs, which means that k must be even. And since there are 21 knights and k must be even, the person who was left out must have been a knight.

We now know that out of the 2020 people who were paired, 2000 were knaves and 20 were knights. Additionally, there were 10 knight-knave pairs. This leaves 1990 knaves who must have been paired with each other, making the number of knave-knave pairs $1990 \div 2 = 995$.

29 (A) 1

The key observation is that a team cannot play in two matches at the same time. Using this observation, we can fill in the remaining matches one by one. A-D cannot take place in rounds 1, 2, 3, or 5, since either A or D is playing in those rounds. So, it must take place in round 4. Now, since A is playing in rounds 1, 3, 4, and 5, A-F must take place in round 2. Since C or E is playing in rounds 2, 3, 4, and 5, C-E must take place in round 1. Similarly, D-E must take place in round 5. Now, B-D must take place in round 3. B-C must take place in round 4. C-F must take place in round 3. And finally, D-F must take place in round 1. We can fill in B-E in round 2 and B-F in round 5 to complete the schedule and prove that it works.

30 (C) 6

Label the vertices A, B, C, and D, starting at the upper left-hand corner and going clockwise. Additionally, let $[\triangle XYZ]$ denote the area of $\triangle XYZ$ for any three vertices X, Y, and Z.

Notice that by the $\frac{1}{2}$ base · height formula for the area of a triangle, $[\triangle DKT] = 2[\triangle TKC]$, $[\triangle DKW] = 2[\triangle WKA]$, $[\triangle BKS] = 2[\triangle SKC]$, and $[\triangle BKP] = 2[\triangle PKA]$.
Therefore, $[\triangle DKT] + [\triangle DKW] + [\triangle BKS] + [\triangle BKP] = 2 \cdot ([\triangle TKC] + [\triangle WKA] + [\triangle SKC] + [\triangle PKA])$.
Equivalently, $([\triangle DKT] + [\triangle DKW]) + ([\triangle BKS] + [\triangle BKP]) = 2 \cdot (([\triangle TKC] + [\triangle SKC]) + ([\triangle WKA] + [\triangle PKA]))$. Or, $18 + 10 = 2 \cdot ([\triangle TKC] + [\triangle SKC] + 8)$. Solving, we get that the area of the shaded quadrilateral is $[\triangle TKC] + [\triangle SKC] = (28 \div 2) - 8 = 6$.

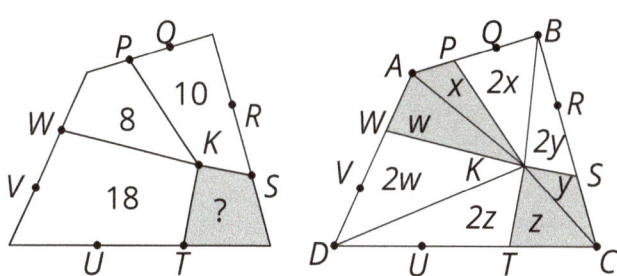

2023

2023

3 Point Solutions

1 (E)

Option (E) is the only configuration of line segments that fits into the given incomplete configuration.

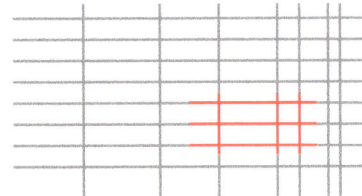

2 (A) triangle

A triangle can be divided into at most one trapezoid (a line from a vertex to the opposite edge yields two triangles, while a line from one edge to another yields one triangle and one trapezoid). The other shapes can all be divided into two trapezoids.

3 (A) 4 or 12

The two holes show numbers which are four hours apart. Therefore, if 8 appears in the first hole, the second will show 12. If 8 appears in the second hole, the first hole will show 4.

4 (B) 12

Observe that the sum of the two numbers in any two opposite edges is the same. For example the edges with end points *a*, *b* and *c*, *d*, respectively, have sum $(a + b) + (c + d) = a + b + c + d$ which is the same as the sum $(a + d) + (b + c)$ of two other opposite edges. So, in our case, the number at the edge with the question mark is $8 + 13 - 9 = 12$.

5 (C)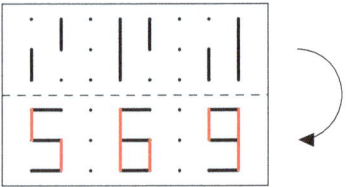

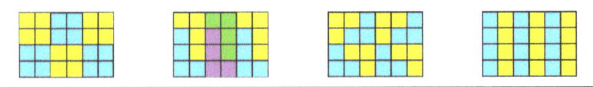

6 (D)

As $4 \times 6 = 24$ is not a multiple of 5, tile (D) cannot be used because it has 5 squares. It is easy to see that all the other tiles can cover the floor.

7 (B) 15

John needs to have 50% of the coins show heads and 50% show tails. He therefore needs to turn 10% of the coins from showing tails to showing heads, and 10% of 150 is 15.

8 (B) *A* and *C*

We can place a second arrow at the end of each of the given arrows to see the bumper cars' paths after 10 seconds have elapsed. These additional arrows are drawn in red in the diagram below. From this we see that cars *A* and *C* will collide.

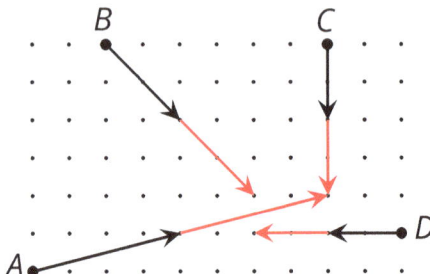

9 (D) 10

Let D_1, D_2, D_3, D_4, and D_5 be the various discs. D_1 has the smallest radius and D_5 the largest. If Anna's tower has D_5 at the bottom, there are 6 options for making a tower:
D_5, D_4, D_3; D_5, D_4, D_2; D_5, D_4, D_1; D_5, D_3, D_2; D_5, D_3, D_1; and D_5, D_2, D_1.

If the tower has D_4 at the bottom, there are 3 possible towers: D_4, D_3, D_2; D_4, D_3, D_1; and D_4, D_2, D_1.

Finally, she can make a tower with D_3 on the bottom, with the only possibility being D_3, D_2, D_1.

In total, this gives 6 + 3 + 1 = 10 towers.

10 (E) 7

The sum of all the whole numbers from 1 to 8 is 1 + 2 + 3 + 4 + 5 + 6 + 7 + 8 = (1 + 8) + (2 + 7) + (3 + 6) + (4 + 5) = 4 × 9 = 36. Since the sum of the numbers in each row must be equal, each row must add up to half of 36, which is 18. Similarly, each column must sum to 9 since we have four columns. Filling out the grid in this way gives

6	4	1	7
3	5	8	2

4 Point Solutions

11 (E) ♡△♡

Notice that the hundreds digit is different in the first and second consecutive whole numbers. This means that in the first number, the tens and ones digits must be 99, and in the second number, they must be 00. Hence ◇ = 9 and △ = 0. But comparing the second and third numbers we must have that □ = 1. The numbers written are thus 199, 200, 201, so we conclude that ♡ = 2. Therefore the next number is 202 = ♡△♡.

12 (C) 18

Let r be the common radius of the semicircles. Consider the line segment that begins at the left side of the left-most semicircle and extends to the right side of the right-most semicircle. Looking along the top semicircles, the length of this segment is $2r + 12 + 2r + 12 + 2r = 6r + 24$. Looking along the bottom semicircles, this segment has length $22 + 2r + 16 + 2r + 22 = 4r + 60$. Equating these expressions gives us the equation $6r + 24 = 4r + 60$. Solving for r yields $r = 18$.

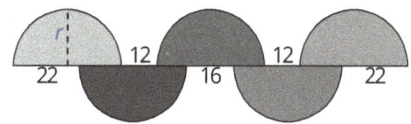

13 (B) 3

Edges of a cube are formed where two faces meet, so each edge is the edge of two faces of the cube. If we color only two edges of the cube red, then at most four of the six faces will have a red edge, which is not enough. We will need at least three red edges. The diagram below gives an example of such an edge coloring:

14 (C) 6

We proceed by considering the number of digits. Among the one-digit positive integers, only 6 and 9 use six matchsticks (we do not count 0 since it is not positive, even though it uses six matchsticks). For a number with two digits, either both digits use three matchsticks, or one uses two and the other uses four, so that the total number of matchsticks will be six. In the first case, we have 14 and 41, and in the second case, we have 77. For a number with three digits, we must have each of the digits using two matchsticks (no number uses only one matchstick). The only number satisfying this is 111. Thus, the six numbers are 6, 9, 14, 41, 77, and 111.

---

15 (E) 12

If we pick vertices that share a side of the square, we can draw an equilateral triangle using the line segment between the vertices as one side, and the last vertex of this triangle is a point which is equidistant from our original vertices. There will be two such triangles for each pair of points, one a reflection of the other. There are four such pairs of points, so we get 2 × 4 = 8 points that are equidistant from two vertices on the square that share a side. If we instead pick vertices along the diagonal of the square, the two points which are equidistant from these points are the other two vertices of the square. This gives an additional four points, for a total of 12 points. This diagram shows all such points:

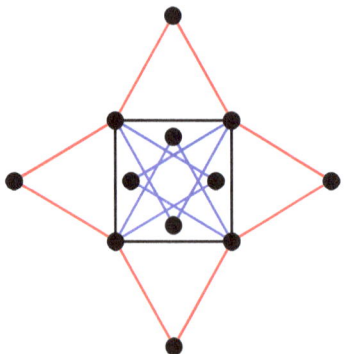

16 (D) 70°

Since △ABC is isosceles, ∠BAC = ∠ACB = (180° − 40°) ÷ 2 = 70°. Considering the triangle △ACF, we have ∠AFC + ∠CAF + ∠ACF = 180°. Solving for the first angle yields
 ∠AFC = 180° − ∠CAF − ∠ACF
   = 180° − (70° − ∠BAE) − ∠ACF.
Using the fact that ∠BAE = ∠ACF, we have ∠AFC = 180° − 70°+ ∠ACF − ∠ACF = 110°. The angle ∠? is supplementary to ∠AFC, so ∠? = 70°.

---

17 (D) 40

Tom and John hit the outer ring a combined total of four times, the second ring four times, and the middle four times. Lily hit each ring twice, so the combined score of Tom and John is exactly twice as much as Lily's score. Tom and John have a combined score of 46 + 34 = 80, so Lily scored 40 points.

---

18 (D) 150

Let h and b be the height and base length of the white rectangle. Then, $\frac{h}{2}$ is the common height of each of the four non-overlapping white triangles and $2b$ is the combined length of the four corresponding bases. Hence, the combined area of the four triangles is $\frac{1}{2} \times 2b \times \frac{h}{2} = \frac{1}{2} \times b \times h$, which is half the area of the white rectangle. The other half is the area of the gray rectangle, and we know that this area is 3 × 25 cm² = 75 cm². So, the area of the whole rectangle is 2 × 75 cm² = 150 cm².

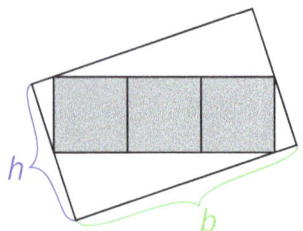

19 (B) 3

Two extra lines are not sufficient to obtain all eight angle values because adding two extra lines gives a total of four lines, and for 4 lines $\ell_1, \ell_2, \ell_3, \ell_4$, there can be at most 6 different angles:

$\angle(\ell_1, \ell_2), \angle(\ell_1, \ell_3), \angle(\ell_1, \ell_4), \angle(\ell_2, \ell_3), \angle(\ell_2, \ell_4)$, and $\angle(\ell_3, \ell_4)$.

Three extra lines suffice, as in the example below.

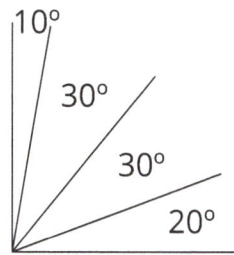

---

20 (A) 4

If we only use positive integers, the sum would be much larger than 2023. Indeed, the only way to sum 2023 positive integers and get 2023 is if all the integers are equal to 1, but we need consecutive integers, which are all distinct. Therefore, we must use negative integers. We can obtain 0 as a sum by taking the consecutive integers −1011, ..., −2, −1, 0, 1, 2, ..., 1011. We can then shift this sequence by one in the positive direction to get our desired sum of 2023. The sequence −1010, ..., 1012 consists of 2023 consecutive integers, and the sum of these is 1011 + 1012 = 2023, since −1010, ..., 1010 sum to 0. The largest of these integers is 1012, and the sum of its digits is 1 + 0 + 1 + 2 = 4.

5 Point Solutions

21 (B) 5

Since three kangaroos are standing next to another kangaroo, it must be the case that all three of these kangaroos are standing next to each other. Otherwise, we would have to break up these three and would have at least one standing alone. So the sequence BKKKB must arise in the circle. Since no two beavers are standing next to each other, we can add a kangaroo to each side of this sequence: KBKKKBK. We cannot add more kangaroos since we already have three standing next to another kangaroo. We can close this circle by adding a third and final beaver. So we have five kangaroos in total.

---

22 (C) $\frac{180}{19}$

The average speed is determined by dividing the total distance traveled by the time taken to travel this distance. Let ℓ be the side length of the equilateral triangle. The total distance traveled by the ant is then 3ℓ cm. To determine the time taken to travel along each edge, we use the formula: distance = (rate)(time), so that time = distance/rate. Thus, the total time taken to traverse all sides of the triangle is $\frac{\ell}{5} + \frac{\ell}{15} + \frac{\ell}{20} = \frac{12\ell + 4\ell + 3\ell}{60} = \frac{19\ell}{60}$ min. This gives an average speed of $\frac{3\ell}{\frac{19\ell}{60}} = \frac{180}{19}$ cm/min.

23 (C) 3

We work backwards through the information given to generate the graph shown below. Since there are seven dwarves, and Doc played six games, he must have played a game against every other dwarf on Monday. Happy played five games, so did not play against one of the dwarves. Happy must not have played against Grumpy, because Grumpy only played one game on Monday, which must have been against Doc. Similarly, Bashful played against all dwarves except for Grumpy and Sneezy, since their games are already accounted for. This completes the graph below (where 1 = Grumpy, 2 = Sneezy, etc.), and we can see that ? = Dopey played three games.

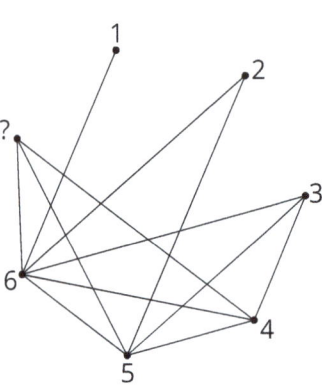

24 (E) 16

The numbers 8 and 9 can only have 1 in an adjacent region, so they must be placed in the square regions on the left and right sides, while 1 is placed in the central hexagon. There are two choices for this placement: 8 on the left and 9 on the right, or vice versa. The numbers 6 and 7 can only have 1 or 2 in an adjacent region, so they must be placed in a region with only two adjacent regions. Therefore, 7 must be placed in one of the four remaining square regions, which gives an additional 4 choices. This choice then determines where 2 is placed (in the triangular region adjacent to 7) and 6 is placed adjacent to 2. At this point, it remains to place 3, 4, and 5. The numbers 4 and 5 cannot be placed in adjacent regions, so they must be put in the two remaining square regions, with 3 in the triangular region between them. There are two choices for this placement: 4 on the left and 5 on the right, or vice versa. In total, this yields 2 × 4 × 2 = 16 configurations, one of which is given below.

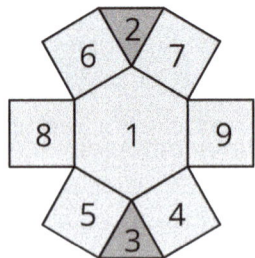

25 (D) 17

Let n be the number of people in front and behind Martin, so he is in the $(n + 1)$th place and there are $2n + 1$ people in line. Martin's friends are behind him.

$$\underbrace{1^{st}, 2^{nd}, 3^{rd}, ..., n^{th}}_{n \text{ people}} \quad \text{Martin} \quad \underbrace{..., 19^{th}, ..., 28^{th}, ..., (2n + 1)^{th}}_{n \text{ people}}$$

28 is an even number and the last position is an odd number, so $28 < 2n + 1$. Also, $n + 1 < 19$. Solving these inequalities yields $13.5 < n < 18$, so n can only be 14, 15, 16, or 17. Then the total number of people in line could be 29, 31, 33, or 35. Since the number of people in line must be a multiple of 3, we have 33 people in line, $n = 16$, and thus Martin is 17th in line.

26 **(B) 11**
Since every mouse left their house, the number of mice that left the two houses on the left is 7 + 8 = 15. The number of mice that arrived at the house on the right is 4. These 4 mice did not use the path of interest, so the number that did is 15 − 4 = 11.

---

27 **(E) 6**
Since the ones digit of 2023 is a 3, we must use a column of nine 7's because the only multiple of 7 ending in a 3 is 7 × 9 = 63. This gives 6 to carry. The sum of the digits in the tens column must end with a 6 so that we have 6 + 6 = 12 to make this end with a 2, which means we need a column of eight 7's as 8 × 7 = 56. Now this column has a 6 to carry, so we need 14 = 7 + 7 on the leftmost column. The whole sum is shown in the figure.

```
   777
   777
    77
    77
    77
 +  77
    77
    77
    77
     7
  ----
  2023
```

---

28 **(A)** $\frac{3}{11}$
Let A denote the area of the larger hexagon and let a denote the area of the smaller hexagon. Our goal is to determine $\frac{a}{A}$. Let S denote the area of the shaded region, so that $\frac{S}{a} = \frac{3}{4}$. Outside of the small hexagon, the shaded and unshaded regions have the same area, so that S is half the area outside of the small hexagon. Therefore $S = \frac{A-a}{2}$. Combining our last two equations, we obtain $\frac{4}{3} = \frac{S}{a} = \frac{A-a}{2a} = \frac{A}{2a} - \frac{1}{2}$. This yeilds $\frac{11}{6} = \frac{A}{2a}$, so that $\frac{a}{A} = \frac{3}{11}$.

---

29 **(A) 18**
We are searching for a sequence of six consecutive whole numbers that we know contains 6, 7, and 8. The possible sequences are (3, 4, 5, 6, 7, 8), (4, 5, 6, 7, 8, 9), (5, 6, 7, 8, 9, 10), and (6, 7, 8, 9, 10, 11). It is impossible to obtain 23 as the sum of three numbers from (3, 4, 5, 6, 7, 8) since the largest possible three-term sum is 21. It is also impossible to obtain 17 from either (5, 6, 7, 8, 9, 10) or (6, 7, 8, 9, 10, 11) since the smallest possible three-term sums are 18 and 21, respectively. So our sequence must be (4, 5, 6, 7, 8, 9), and we are missing the numbers 4, 5, and 9. Their sum is 18.

---

30 **(C) 24**
The average points-per-game among the seventh, eighth, and ninth games is $\frac{24+17+25}{3} = \frac{66}{3}$ = 22. Their average score in the first six games must have been less than 22. Since 22 × 6 = 132, they could have scored at most 131 total points in their first six games. Their average after 10 games was more than 22, so they scored more than 22 × 10 = 220 points over these 10 games (i.e., at least 221 points). In order to score 221 points in total, they need to score 221 − 131 − 66 = 24 points in the tenth game.

2025

2025

3 Point Solutions

1 (C) 5220
To make the largest number possible, Lisa should put the larger digits in higher place values. Therefore the digits 2, 0, 2, 5 must be arranged in decreasing order, meaning that the largest number Lisa can make is 5220.

2 (E) 12
Note that in every rotation, the light gray triangle moves one space clockwise, and there are six total spaces in the hexagon. To end up where it started, it needs to make a number of rotations that is a multiple of 6. The only multiple of 6 in the options is 12.

3 (E) ⚅
If one of the dice is 6, then the other two dice must sum to 2, which means they both must be 1, but then they are not different. All other numbers are achievable: 1 + 2 + 5 = 1 + 3 + 4 = 8.

4 (B) $\frac{1}{3}$
There are 36 triangles total, of which 12 are shaded. Thus the answer is $\frac{12}{36} = \frac{1}{3}$.

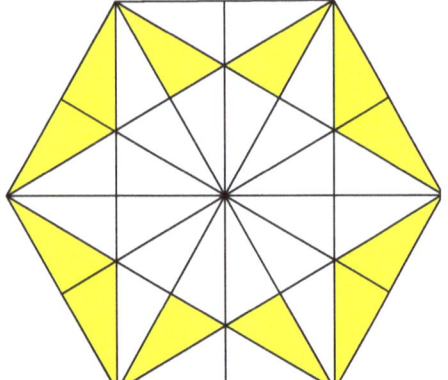

5 (A) 60
There are 60 minutes in one hour, so there are 12 × 60 minutes in 12 hours. The answer is 12 × 60 ÷ 12 = 60.

6 (E) 30
Dominic's age right now is 5 + 6 = 11. Therefore, in 7 years Daniel will be 5 + 7 = 12 and Dominic will be 11 + 7 = 18. The requested sum is 12 + 18 = 30.

7 (C) −5
To make the smallest number possible, Omar should put the larger digits in boxes after a minus sign, so a large value gets taken away from the result. Since there are two subtracted digits, we put the two largest digits there: 5 and 2. The other two boxes are filled with 2 and 0. Therefore, the minimum possible result is (2 + 0) − (5 + 2) = −5.

8 (B) 5
Both truth-tellers and liars would answer "yes" to the question, "Are you a truth-teller?"; truth-tellers tell the truth, so they would truthfully answer "yes." Liars always lie, so although they are not truth-tellers, they would lie and answer "yes." Therefore, since 20 people answered "yes," there are 20 people in the room. Since the truth-tellers have ten more people than the liars, if we take those extra ten truth-tellers away, the two groups would be equal. When we take away those ten people, 10 people remain to be split evenly between the two groups. This means there are 5 liars and 5 + 10 = 15 truth-tellers.

9 (B) 36 cm²

Without overlap, the five circles have a combined area of 5 × 8 = 40 cm². However there are 4 overlaps of size 1 cm², so 4 × 1 = 4 cm² of the area covered by the overlaps. Therefore, removing the overlaps, the circles cover 40 − 4 = 36 cm².

---

10 (A) 4037

Each wheel has the numbers from 0 to 9 in order, so if you look at a digit from the side, the real combination will have that same digit shifted 2 places forward, so we can just add 2 to each digit to find the real combination (wrapping around 8 → 0 and 9 → 1). Therefore, we get 2815 → 4037, so the answer is 4037.

4 Point Solutions

11 (C) 8

In each cell, write the number of ways to get there, as shown on the right. Each number is the sum of the number above it and to its left (since those are the only cells that can reach it). Therefore, the answer is 2 + 4 + 2 = 8.

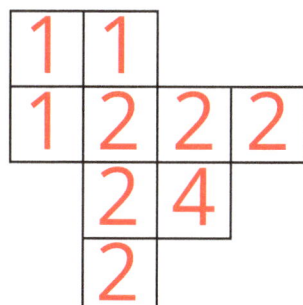

12 (A) 16 m

There are 4 gaps between the 5 hurdles, each 8 m long. Thus, the distance from the first hurdle to the last hurdle is 4 · 8 m = 32 m. The first hurdle is at 12 m, so the total distance from the start to the last hurdle is 12 m + 32 m = 44 m. The race is 60 m long. Therefore, the distance from the last hurdle to the finish line is 60 m − 44 m = 16 m.

---

13 (D) −3

If there are three consecutive circles with two of them filled, then we can fill out the remaining circle (if the middle one is missing, then it is the sum of the numbers on either side; if an edge is missing, then we can take the middle number and subtract the other edge). Doing this, we get −3.

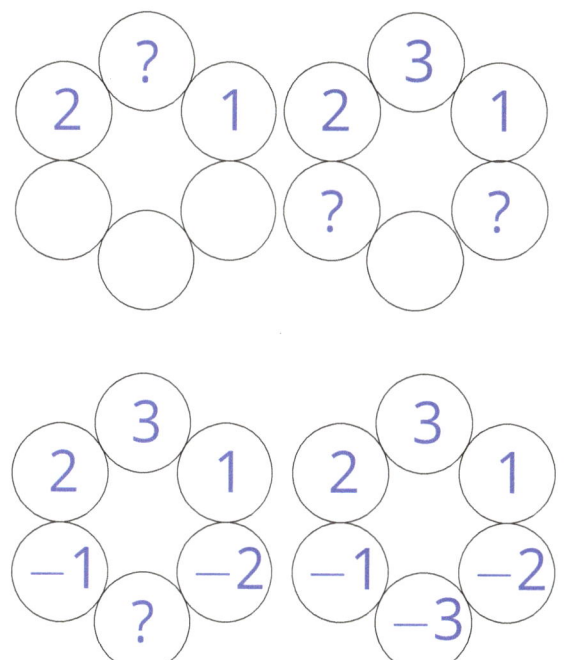

14 (B) 70

We will compute the angles inside the quadrilateral "hole" between the three pictures. Recall that the sum of angles in a quadrilateral is 360°. Recall that along a line, the sum of the angles is 180°; thus, we can compute the bottom-left angle inside the three pictures as 180° − 90° − 62° = 28°. Similarly, the top angle inside the quadrilateral is 90° − x°. Thus, the angles inside the quadrilateral are 28°, 42°, 270° (which is the reflex angle of the corner of the boat picture), and 90° − x°. These should add to 360°, so we have
28 + 42 + 270 + 90 − x = 360 →
x = 28 + 42 = 70.

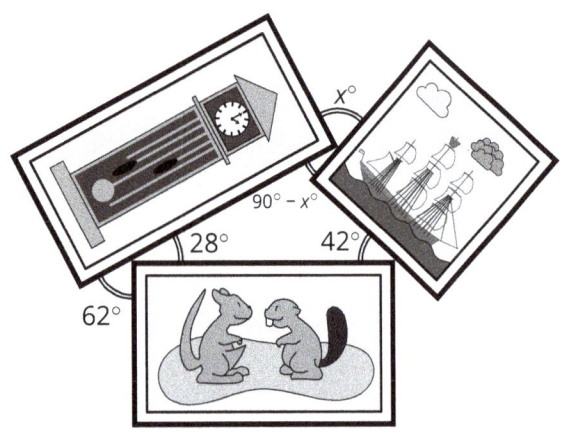

15 (D) 18:15

At any time, the total session time is the sum of the elapsed time and the remaining time. We can see from the given stopwatches that the total time is 14:58 + 21:32 = 36:30. When the two watches show the same time, their total will still be 36:30, so each of them will show 36:30 ÷ 2 = 18:15.

16 (C) 10

There are only 8 primes less than 20: 2, 3, 5, 7, 11, 13, 17, 19. Since there are 8 boxes and they have to all be different, we have to use each of these primes once and only once. In particular, we need to pick 7 of these numbers, and their sum should be a multiple of the remaining number. That is, if S is the sum of all the numbers, and we decide to put the number x in the denominator, then $A = \frac{S-x}{x} \to Ax = S - x \to S = (A + 1)x$. Since A and x are integers, S must be a multiple of both $A + 1$ and x. Here S = 2 + 3 + 5 + 7 + 11 + 13 + 17 + 19 = 77. Therefore x must equal 7 or 11; they both work, and they give $A = \frac{70}{7}$ = 10 and $A = \frac{66}{11}$ = 6 respectively. Therefore the largest possible value of A is 10.

17 (B) 2 circles and 4 crosses

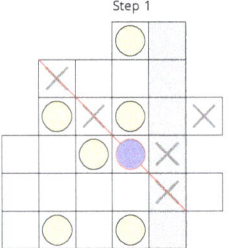
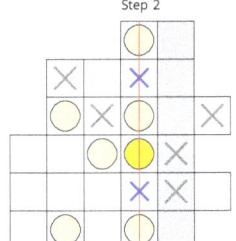
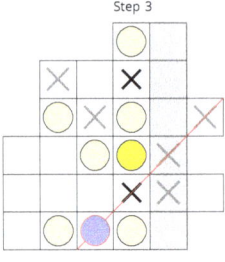
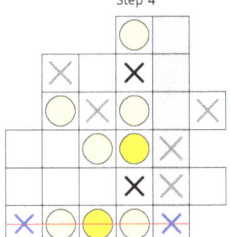
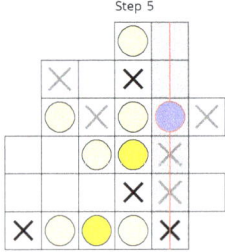
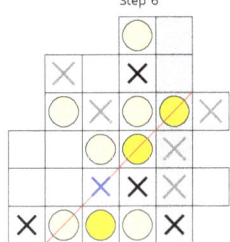
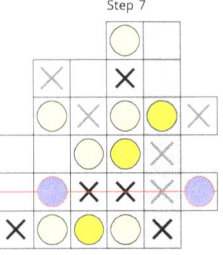
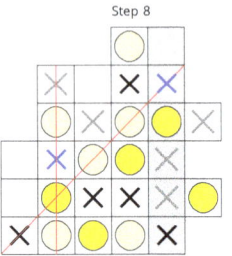
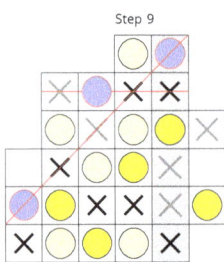

18 (D) 10 cm

Note that $AB + EF = DC + EF = (DE + EF + FC) + EF = (DE + EF) + (FC + EF) = FD + EC$.
Since opposite sides are parallel and each angle of the rectangle is 90°, we have $\angle CBE = \angle CEB = 45°$ and $\angle DFA = \angle DAF = 45°$. In particular, the triangles CBE and DFA are isosceles, with $FD = AD$ and $EC = BC$; since $ABCD$ is a rectangle, $AD = BC$. Thus from the above equation, $AB + EF = FD + EC = BC + BC = 2BC$. Since $AB + EF = 20$ cm, it follows that $BC = 10$ cm.

19 (A) 6

The averages of the two bowls are
$\frac{1+2+6+7+10+11+12}{7} = 7$ and $\frac{3+4+5+8+9}{5} = 5.8$.
Therefore to increase the average in bowl X, we need to remove a ball with value < 7 (since its average is already 7, if we remove a number ≥ 7, its average cannot go up). Similarly, to increase the average in bowl Y, we need to add a ball with value > 5.8. Therefore the moved ball must have value more than 5.8, but less than 7. The only such ball is ball 6.

20 (B) 6 cm

From bottom-left in clockwise order, suppose the quarter circles have radii a, b, c, d, and the length we want to find is x. Looking at the left, top, right, and bottom sides of the rectangle, we get that $a + b = 9$, $b + c = 12$, $c + d = 9$, and $d + x + a = 12$. Adding the first and third equations gives $a + b + c + d = 18$. Adding the second and fourth equations gives $a + b + c + d + x = 24$. Subtracting these yields $x = 6$ cm.

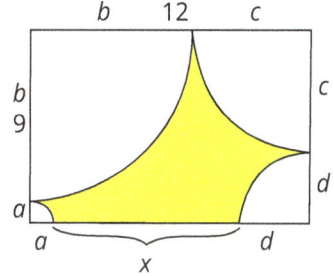

5 Point Solutions

21 (A) 432

We have $Y = 2P = 3A$, so Y is a multiple of 2 and a multiple of 3. Since Y is a digit, the only possible value for Y is 6. Thus $Y = 6$, $P = 6 \div 2 = 3$, and $A = 6 \div 3 = 2$, so the answer is $6 \times 2 \times 3 \times 2 \times 3 \times 2 = 432$.

22 (D) 9

Let x_i be the number of shots in the i-th session. The total number of shots is $x_1 + x_2 = 17$. Writing the percenages as fractions, the total number of shots that hit is $\frac{3}{5}x_1 + \frac{3}{4}x_2$. For this to be an integer, x_1 must be a multiple of 5, and x_2 must be a multiple of 4. Since they are at most 17, the only possible choices for x_1 are 5, 10, 15, which would have $x_2 = 12$, 7, or 2. Since 12 is the only multiple of 4, the only possibility is $(x_1, x_2) = (5, 12)$. The number of shots that hit in the second session is $75\% \cdot 12 = 9$.

23 (E) 16

When Anurag walks, he takes 1 km $\cdot \frac{1}{4}$ h/km $= \frac{1}{4}$ h = 15 min (since an hour is 60 minutes). When Anurag bikes, he takes 1 km $\cdot \frac{1}{15}$ h/km $= \frac{1}{15}$ h = 4 min. Therefore, biking takes 11 less minutes than walking, and since Anurag arrives 5 minutes early by walking, he will arrive $5 + 11 = 16$ minutes early by biking.

24 (C) 66

Complete the rectangle $AEFG$ as shown. It has dimensions $(3 + 4 + 2) \times (3 + 4 + 4 + 2) = 9 \times 13$. We can compute $AG = 3 + 4 + 2 = 9$, $GB = 3 + 4 - 2 = 5$, $BF = 2 + 4 + 2 = 8$, $CF = 2 + 4 = 6$, $EC = 3$, $DE = 4 + 2 - 3 = 3$. The area of the shaded quadrilateral can be found by subtracting the corner triangles from the large rectangle:
$9 \times 13 - \frac{3 \times 3}{2} - \frac{6 \times 8}{2} - \frac{5 \times 9}{2} = 66$

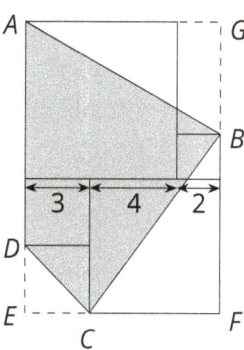

25. (C) 34

Since $p + q = 69$ and p and q do not differ by more than 4, p and q must be close to their average, which is $69 \div 2 = 34.5$. Thus the only possibilities for p and q are (34, 35) or (33, 36); if we decrease the first number further, the second number is more than 4 away.

Similarly, for $r + s = 72$, the only possibilities are (35, 37) and (34, 38).

Note that if p and q are (34, 35), then neither possibility for r and s will work out, since the first uses 35 and the second uses 34 (but all numbers must be distinct). Thus p and q are 33 and 36. Given this, we cannot have r and s be 34 and 38, since 33 and 38 are too far apart (their difference is 5, so they can't be part of a list of 5 consecutive numbers). Thus, r and s are 35 and 37.

Therefore, the numbers already used are 33, 36, 35, and 37. The missing number is 34.

---

26. (D) 200

Let s be the side length of the cube after reducing its height by 3. The surface area that is lost during the removal are the four lateral rectangles of size $s \times 3$, since the top $s \times s$ face of the cuboid is removed, but an equally-sized $s \times s$ face is revealed after the removal (on the bottom of the removed piece). Therefore, the total surface area lost is $4 \times (s \times 3) = 12s = 60$ cm². Thus, $s = 5$ and the original cuboid has volume $8 \times 5 \times 5 = 200$ (cm³).

---

27. (A) 13

Draw the diagonal AC. Since triangles ANB and ANC have the same height (the distance from A to BC), the ratio of their areas is equal to the ratio of their bases (the area is $A = \frac{bh}{2}$, and if the height is the same, the area is proportional to the length of the base). Since $NB = 2NC$, the area of ANB is twice ANC. Thus the area of ANC is $6 \div 2 = 3$.
We do the same for ACK and DCK: they have the same height (the distance from C to AD) and the same base ($AK = D$), so they have equal area: 2. Thus the area of $ABCD$ is $6 + 3 + 2 + 2 = 11$.

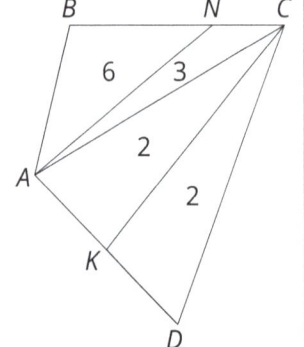

28. (A) 27

Number the wires from top to bottom 1, 2, 3, 4. Since there are more birds below Nha than Trang, Nha is on a higher wire than Trang (and there are 3 birds between them). Also, Trang cannot be on wire 4, since there are birds below her.

Thus, there are three possibilities:
• Nha is on wire 1, and Trang is on wire 2: there are 3 birds on wire 2 (including Trang), and 5 birds below it. However, this means wire 1 would have to have at least $25 - 3 - 5 = 17$ birds for Long's condition to work out, but then it is now impossible for Ha to have 10 birds above her.
• Nha is on wire 1, and Trang is on wire 3: there are 3 birds on wire 2 and 3 (including Trang). Then, there are 2 birds on wire 4, and for Long's condition to work out we must have at least $25 - 3 - 2 = 20$ birds on wire 1. However, it is now impossible for Ha to have 10 birds above her.
• Nha is on wire 2, and Trang is on wire 3: there are 3 birds on wire 3 (including Trang). Then, there are 2 birds on wire 4, and for Long's condition to work out we must have at least $25 - 2 - 3 = 20$ birds on wires 1 and 2. In particular, Long can now be on wire 2 and have 10 birds on wire 1; that is, from top to bottom, the wires have [10, 12, 3, 2] birds, and Ha is on wire 2, Long is on wire 4, Nha is on wire 2, and Trang is on wire 3. The total number of birds is $10 + 12 + 3 + 2 = 27$.

29 (C) Only light gray.

Label the vertices of the octahedron as shown.

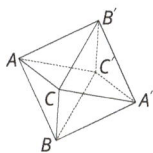

Vertices with the same letter (like A and A') are opposites and so will be colored with the same color. If we label the leftmost point of the colored triangle as A, we get the following net. To get the missing letters, note that whenever we have two triangular faces that share an edge, two of the vertices stay the same, but the third becomes its opposite point (e.g., A ↔ A').

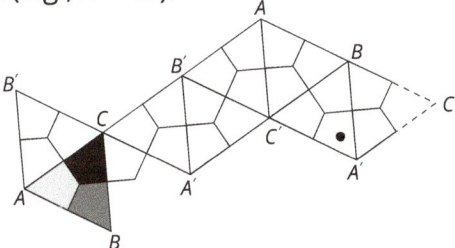

Since the dotted part is adjacent to A', its color will be the same as A, which is light gray. The full net is shown below.

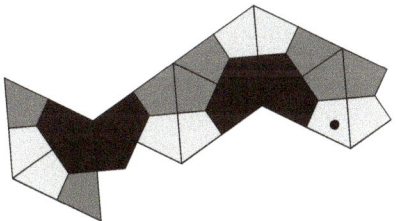

30 (D)

No matter what we see in box (D), we can always deduce where the golden pearls are:
• If box (D) has the golden pearls, we are done.
• If box (D) has the red pearls, then box (A) must have the golden pearls.
• If box (D) has the pink pearls, then box (B) has the black pearls, so box (C) must have the golden pearls.
• If box (D) has the white pearls, then box (E) must have the pink pearls, so box (B) must have the black pearls and box (C) must have the golden pearls.

Part III
Answer Key

	2007	2009	2011	2013	2015
1	C	A	D	D	A
2	A	C	A	D	A
3	E	C	B	A	B
4	D	B	A	C	E
5	D	D	C	E	A
6	D	D	C	E	D
7	C	C	E	C	D
8	A	C	C	E	C
9	B	C	C	A	C
10	C	B	B	C	D
11	C	B	C	E	B
12	C	A	E	E	C
13	D	C	C	B	B
14	E	A	B	C	D
15	C	D	A	B	E
16	D	E	A	A	C
17	A	A	E	A	C
18	B	A	C	E	B
19	B	E	B	B	A
20	E	D	C	C	B
21	D	B	D	B	C
22	D	C	E	D	A
23	E	C	B	C	B
24	B	D	D	C	C
25	A	C	D	B	C
26	C	B	B	A	D
27	C	D	A	B	D
28	D	D	B	D	E
29	D	D	B	B	D
30	D	C	B	C	D

	2017	2019	2021	2023	2025
1	B	E	A	E	C
2	C	E	E	A	E
3	C	C	D	A	E
4	A	D	B	B	B
5	B	D	A	C	A
6	E	D	A	D	E
7	D	A	B	B	C
8	D	B	E	B	B
9	B	D	B	D	B
10	A	A	E	E	A
11	E	A	D	E	C
12	A	B	E	C	A
13	D	E	C	B	D
14	A	B	B	C	B
15	E	C	D	E	D
16	A	C	C	D	C
17	B	D	B	D	B
18	B	B	C	D	D
19	B	C	E	B	A
20	D	C	C	A	B
21	C	B	C	B	A
22	C	B	D	C	D
23	B	C	B	C	E
24	A	E	B	E	C
25	D	E	D	D	C
26	A	A	B	B	D
27	D	C	C	E	A
28	E	E	D	A	A
29	D	E	A	A	C
30	D	D	C	C	D

www.ingramcontent.com/pod-product-compliance
Lightning Source LLC
Chambersburg PA
CBHW061212230426
43665CB00032B/2991